Lecture Notes in Computer Science   **8970**

Commenced Publication in 1973
Founding and Former Series Editors:
Gerhard Goos, Juris Hartmanis, and Jan van Leeuwen

More information about this series at http://www.springer.com/series/8637

Abdelkader Hameurlain · Josef Küng
Roland Wagner · Ladjel Bellatreche
Mukesh Mohania (Eds.)

Transactions on Large-Scale Data- and Knowledge- Centered Systems XVII

Selected Papers from DaWaK 2013

 Springer

Editors-in-Chief
Abdelkader Hameurlain
IRIT, Paul Sabatier University
Toulouse
France

Roland Wagner
FAW, University of Linz
Linz
Austria

Josef Küng
FAW, University of Linz
Linz
Austria

Guest Editors
Ladjel Bellatreche
LIAS/ISAE-ENSMA
Chasseneuil-du-Poitou
France

Mukesh Mohania
IBM India Research Lab
New Delhi
India

ISSN 0302-9743   ISSN 1611-3349   (electronic)
Lecture Notes in Computer Science
ISBN 978-3-662-46334-5   ISBN 978-3-662-46335-2   (eBook)
DOI 10.1007/978-3-662-46335-2

Library of Congress Control Number: 2015930728

Springer Heidelberg New York Dordrecht London

Printed on acid-free paper

Springer-Verlag GmbH Berlin Heidelberg is part of Springer Science+Business Media
(www.springer.com)

Special Issue of DaWak 2013

We welcome you to this special issue dedicated to the best papers presented at the 15th International Conference on Data Warehousing and Knowledge Discovery (DaWaK) that was held in Prague, Czech Republic, during August 26–29, 2013. Data Warehousing and Knowledge Discovery has been widely accepted as a key technology for enterprises and organizations to improve their abilities in data analysis, decision support, and the automatic extraction of knowledge from data. With the exponentially growing amount of information to be included in the decision-making process, the data to be considered becomes more and more complex in both structure and semantics. New developments such as cloud computing and Big Data add to the challenges with massive scaling, a new computing infrastructure, and new types of data. Consequently, the process of retrieval and knowledge discovery from this huge amount of heterogeneous complex data builds the litmus-test for the research in the area. During the past years, the International Conference on Data Warehousing and Knowledge Discovery (DaWaK) has become one of the most important international scientific events to bring together researchers, developers, and practitioners to discuss the latest research issues and experiences in developing and deploying data warehousing and knowledge discovery systems, applications, and solutions. DaWaK is in the top 20 of the google scholar ranking related to Data Mining and Analysis: http://scholar.google.com/citations?view_op=top_venues&hl=fr&vq=eng_datamininganalysis.

The DaWaK 2013 call for papers attracted 89 papers and the Program Committee finally selected 24 full papers and 8 short papers, making an acceptance rate of 36% of submitted papers. The accepted papers cover a number of broad research areas on both theoretical and practical aspects of data warehouse and knowledge discovery. In the area of data warehousing, the topics covered included conceptual design, query optimization, MapReduce paradigm, scalability, data compression, materialized views, data partitioning, distributed and parallel processing and data warehouses and data mining applications integration, recommendation and personalization, multidimensional analysis of text documents, and data warehousing for real-world applications such as health, spatial applications, energy, etc. In the areas of data mining and knowledge discovery, the topics included stream data analysis and mining, dimensionality reduction, and traditional data mining techniques topics such as frequent item sets, clustering, association, classification ranking, and application of data mining technologies to real-world problems. It is especially notable to see that some papers covered emerging real-world applications as bioinformatics, social network, mobile data, energy power, email management, environment surveillance as well as integration of multiple technologies such as conceptual modeling, evaluation metrics, and OLAP mining.

Out of the 24 full papers, we selected six papers to be invited for the special issue in the Journal LNCS Transactions on Large-Scale Data- and Knowledge-Centered Systems and after a second round of review we finally accepted five papers. Thus, the relative acceptance rate for the papers included in this special issue is competitive. Needless to say, these five papers represent innovative and high-quality research, where two papers cover data warehousing aspects related to query processing optimization in advanced platforms (MapReduce and parallel databases) and three cover knowledge discovery (causal network inference problem, dimensionality reduction, and the quality of pattern mining task). The particularity of these papers is that most of them used case studies issued from international projects (e.g., EU-funded FP7) and real systems. We congratulate the authors of these five papers and thank all authors who submitted articles to DaWaK.

The five selected papers are summarized as follows:

The first paper, titled Data Warehouse Processing Scale-up for Massive Concurrent Queries with SPIN, by João Costa and Pedro Furtado, presents a data processing model to handle large concurrent workloads executed on the top of a data warehouse. In this situation, queries compete for the same resources and consequently, their response time may significantly increase. The proposal is based on a solid theory. The most important state-of-the-art studies are analyzed and the authors show their inefficiency in terms of scalability and large memory requirements. The proposed model called SPIN is developed to predict execution times when a large set of aggregated star queries are concurrently executed. SPIN is accompanied with operators like selection, projection, computation, data switches, and aggregation operators. These operators are processed in the pipeline way. The functional architecture of SPIN system is given and evaluated using TPC-H benchmark.

The second paper, titled An Uncoupled Data Process and Transfer Model for MapReduce, by Li Zha, Jie Zhang, Wei Liu, and Jian Lin, proposes an approach to optimize the functions of MapReduce in order to improve the performance of a system running in a cloud environment. The proposed model uncouples the dependency relationship of maps and reduces in the original MapReduce in order to make full use of the network bandwidth in map and reduce tasks and balance the network load. Issues regarding workload balancing, data transferring, and fault tolerance are studied in the proposed model. The proposal is evaluated in Baidu, the biggest search engine company in China, to evaluate its efficiency and effectiveness. The proposed model is transparent to end users and compatible with the original Hadoop. Efficient state-of-the-art methods improve the throughput of a given system by making full use of its resources. We can cite, for instance, Hadoop Online Prototype, Copy-compute Splitting, Dynamic Weight Assignment, MapReduce Energy Efficiency, etc.

The third paper, titled Enhanced Fast Causal Network Inference over Event Streams, by Saurav Acharya and Byung Suk Lee, addresses a challenging problem of causal network inference in the context of streaming events. Several advanced applications are concerned by causal inference in diverse domains such as health care, stock markets, smart electric grids, and social media. The main issue of these applications is to infer the causes of abnormal activities immediately from their event streams. To deal with this problem, two algorithms are presented: OATNI for Order-Aware Temporal Network Inference and EFCNI for Enhanced Fast Causal Network Inference. The OATNI exploits a temporal network structure (a directed acyclic graph) to represent temporal precedence relationships between event types. The EFCNI uses a time-centric causal modeling strategy to speed up the learning of causal network and to learn an accurate causal network even if the streaming events are out of order. Complexity analysis of the proposed algorithms is given. These algorithms are compared theoretically and experimentally against two state-of-the-art algorithms, PC and Fast Causal Network Inference. These experiments show their efficiency in terms of computation time and resiliency to out-of-order in streams.

The fourth paper, titled Learning through Non-linearly Supervised Dimensionality Reduction, by Josif Grabocka and Lars Schmidt-Thieme proposes a novel dimensionality reduction which simultaneously reconstructs the predictors by the means of matrix factorization and estimates the target variable. The originality of this proposal is the use of a nonlinear SVM as the classification loss term. A consistent state of the art has been established that includes: Dimensionality Reduction, Matrix Factorization, and Supervised Dimensionality Reduction. The proposed algorithm called Nonlinearly Supervised Dimensionality Reduction (NSDR) is compared against three algorithms, PCA-SVMs (Principal Component Analysis and then SVMs classification), SVMs, and LSDR (Linearly Supervised Dimensionality Reduction) using five real-life datasets. The obtained results show a significant improvement against unsupervised techniques. Application of this work is developed in two FP7 projects: Reduction (www.reduction-project.eu) and iTalk2Learn (www.italk2learn.eu).

The fifth paper, titled Metrics for Association Rule Clustering Assessment, by Veronica Carvalho, Fabiano Santos, and Solange Rezende, addresses a difficult problem that consists in evaluating a pattern mining task. The authors consider the case of association rule clustering. A questionnaire is used to understand the important aspects to be considered by a user when association rule mining is preceded by clustering. Benefiting from these results, 11 metrics are proposed to assess the quality of clustering-based association rule mining. Basically, each metric compares the set of mined association rules with the set that would be mined without clustering. This work provides relevant metrics for evaluating the quality pattern mining algorithms.

January 2015   Ladjel Bellatreche
                                                                            Mukesh Mohania

Organization

Program Committee

Alberto Abelló	Universitat Politècnica de Catalunya, Spain
Mohammed Al-Kateb	Teradata Labs, USA
Ladjel Bellatreche	LIAS/ISAE-ENSMA, France
Petr Berka	University of Economics, Prague, Czech Republic
Vasudha Bhatnagar	University of Delhi, India
Karen Davis	University of Cincinnati, USA
Dejing Dou	University of Oregon, USA
Selma Khouri	National High School for Computer Science, Algeria
Sofian Maabout	University of Bordeaux, France
Mukesh Mohania	IBM Research, India
Lu Qin	University of Technology, Sydney, Australia
Arnaud Soulet	University of Tours, France
Anand Prabhu Subramanian	IBM Research, India
Olivier Teste	IRIT, France
Panos Vassiliadis	University of Ioannina, Greece

Editorial Board

Reviewers

Contents

Data Warehouse Processing Scale-Up
for Massive Concurrent Queries with SPIN

João Pedro Costa[1,2(✉)] and Pedro Furtado[2]

[1] DEIS, ISEC, Polytechnic Institute of Coimbra, Coimbra, Portugal
jcosta@isec.pt
[2] University of Coimbra, Coimbra, Portugal
pnf@dei.uc.pt

Abstract. Data Warehouses (DW) store valuable information not only for strategic business decisions, but also for operational daily decisions. As a consequence, a large number of queries are concurrently submitted, stressing the database engine ability to handle such query workloads without severely degrading query response times. The query-at-time model of common database engines, where each query is independently executed and competes for the same resources, is inefficient for handling large DWs and does not provides the expected performance and scalability when processing large numbers of concurrent queries. Related work shows that there's a performance advantage on sharing data and processing, but the proposed solutions suffer from memory limitations, reduced scalability and unpredictable execution times when applied to large DWs, particularly those with large dimensions. SPIN proposes an approach to share computation and data among concurrent queries that delivers scale-up, even in the presence of massive query workloads. In this paper we describe the mechanisms used by SPIN to embed data and queries into a shared query processing pipeline tree and how SPIN dynamically reorganizes the processing tree. We also provide experimental validation of the approach.

1 Introduction

With the query-at-a-time model of common database systems, each query competes for resources (IO, CPU, …) and is independently executed without any processing and data sharing considerations. Relations are concurrently scanned by each query of the concurrently running workload, and tuples are independently filtered, joined and aggregated. While this may not raise performance issues for most operational systems, it is a performance killer when dealing with large Data Warehouses (DW), with large fact and dimension relations. Therefore, the database engine is unable to handle concurrent query workloads scalably without significantly affecting the query execution time. Consequently, predictable execution times under scalable data volumes and query workloads can only be attained through high level data and processing sharing among concurrently running queries.

Recent proposals aimed to provide improved sharing [1] focus on sharing fact table reading and joining costs among running queries, by using a set of dimension filters to perform fact-to-dimension joins. However, the usefulness of those approaches is

© Springer-Verlag Berlin Heidelberg 2015
A. Hameurlain et al. (Eds.): TLDKS XVII, LNCS 8970, pp. 1–23, 2015.
DOI: 10.1007/978-3-662-46335-2_1

limited to small dimensions that can fit entirely in memory, as recognized in [2], therefore large dimensions may severely degrade performance.

We argue that there's a need for improved data and processing sharing among a large number of concurrently running star queries. Such approach should also deal with scalable data volumes and processing infrastructures.

Our SPIN approach is a data and processing sharing model that can deliver predictable execution times to a large set of concurrently running aggregation star queries. It has minimum memory requirements and can handle large data volumes and be deployed over scalable processing infrastructures with almost linear speedups. We describe the mechanisms used by SPIN to embed data and queries into a shared query processing pipeline tree and how SPIN dynamically reorganizes the processing tree. We discuss how SPIN characteristics overcome the limitations of recent proposals on data and processing sharing, such as memory limitations, reduced scalability and unpredictable execution times when applied to large DWs, particularly those with large dimensions.

The paper is organized as follow: Sect. 2 reviews related work on data and processing sharing, and their limitations on delivering scalable and predictable performance. Section 3 presents and discusses SPIN and how it can overcome such limitations and Sect. 4 presents implementation details of the SPIN prototype. We evaluate SPIN in Sect. 5, and finally we present conclusions in Sect. 6.

2   Related Work

The usage pattern of DWs is changing from the traditional, limited set of simultaneous users and queries, mainly well-known reporting queries, to a more dynamic and concurrent environment, with more simultaneous users and ad hoc queries. DW query patterns are mainly composed by star aggregation queries, which contain a set of query predicates (filters) and aggregations. Figure 1 illustrates the query template.

```
SELECT dim attributes, aggregation functions
FROM   fact, set of dimension tables
WHERE  join conditions
AND    dim attribute conditions
GROUP BY dim attributes
```

Fig. 1. Template of an Aggregated Star Query

The query-at-a-time execution model of traditional RDBMS systems, where each query is executed independently, does not provide a scalable environment to handle much larger, concurrent and unpredictable workloads. The use of a parallel infrastructure does not solve this problem because the additional computational and storage capabilities only lessen it, while introducing others problems such as load-balancing, optimal data distribution and network capacity.

Queries submitted to a star schema model have common processing tasks, particularly those related to IO processing of the fact table (costly operations).

Queries submitted to a star schema model have common processing tasks, particularly those related to IO processing of the fact table (costly operations). Analyzing the execution query plan, we observe that the low-level data access methods, such as sequential scan, represent a major weight in the overall query execution time. One way to reduce such a burden is to store relations in memory. However, the size of the physical memory may be insufficient to hold large DW, and at the same time for performing join and sort operations.

Cooperative scans [3] enhances performance by improving data sharing between concurrent queries, by dynamic scheduling queries and their data requests according to the current executing actions. While this minimizes the overall IO costs, by mainly using sequential scans instead of a large number of costly random IO operations, and the number of scan operations (since scans are shared between queries), it introduces undesirable delays to query execution, since the execution of some actions may have to be postponed, and does not deliver predictable query execution times.

QPipe [4] applies on-demand simultaneous pipelining of common intermediate results across queries, avoiding costly materializations and improving performance when compared to tuple-by-tuple evaluation. Each operator is promoted to an independent micro-engine, called μEngine, which accepts request and serve them as queues. It introduced the concept of Window of opportunity, as the time interval where newly submitted operators can take advantage of the one already in progress. Resource utilization is improved when requests of the same nature are grouped together, and when dedicated processes are used to process each group of similar requests.

Crescando [5] is based on parallel and collaborative scans in main memory and the so-called "query-data" joins known from data-stream processing. Crescando loads a tuple into memory and then "joins" the tuple with all interested queries, so that the cost associated with loading the tuple into memory is amortized. While the proposed approach is not always optimal for a given workload, it provides latency and freshness guarantees for all workloads.

DataPath [6] is a "data-centric" system where queries do not request the data, instead the data is automatically pushed onto processors. It resembles the QPipe and the main-memory-based Crescando system in the way it attempts to share memory access latency and bandwidth.

SharedDB [7] introduce the concept of Global Query Plans, which compiles a single plan for the whole workload, instead of compiling each individual query into separate plans. This plan serves multiple concurrent queries and may be reused over a long period of time. The proposed Shared Join plans approach, which combines (union) relation tuples of all concurrent queries before performing a single large shared join, instead of multiple smaller joins, only proved to be efficient for large number of concurrent queries. SharedDB batches queries and updates and, thereby making use of traditional, best-of-breed algorithms to implement joins, sorting, and grouping. While one batch of queries and updates is processed, newly arriving queries and updates are queued. When the current batch of queries and updates has been processed, then the queues are emptied in order to form the next batch of queries and updates. This batch-based execution model adds latency to each query. A specific advantage of SharedDB as compared to QPipe and DataPath is its ability to meet SLAs and bound the response time of queries.

CJOIN [1, 2] applies a continuous scan model to the fact table, reading and placing fact tuples in a pipeline, and sharing dimension join tasks among queries, by attaching a bitmap tag to each fact tuple, one bit for each query, and attaching a similar bitmap tab to each dimension tuple referenced by at least one of the running queries. Each fact tuple in the pipeline goes through a set of filters (one for each dimension) to determine if it is referenced by at least one of the running queries. It not, the tuple is discarded. Tuples that reach the end of the pipeline (tuples not discarded in filters) are them distributed to dedicated query aggregations operators, one for each query.

CJoin overcomes the limitations of the query-at-a-time model, allowing high level of concurrency, with multiple concurrent queries being processed at a time, by scheduling processing tasks so that they can share IO, particularly scanning tasks. In this model, after creating the execution plan of multiple queries a pre-processor analysis the processing tasks and schedules them so that they can share processing tasks. This is applied not only to IO processing tasks but also to filtering and aggregation tasks, arranged in a pipelined fashion. The system is continuously scanning the fact table and tuples are put in a pipeline for processing. The pre-processor add a bit vector to each tuple that it receives, one bit for each query in the workload, before forwarding the tuples to the pipeline. A similar bit vector is also added to each dimension tuple, where each bit indicates if the dimension tuple satisfies the restrictions (of filtering conditions) of the corresponding query. This bit vector information is used to decide (filter) which fact tuples satisfy the conditions and should be forward into the pipeline.

While this approach reduces IO cost, it requires all dimension tables to reside in memory in order to be probed for performing hash joins, and to continuously update dimension bit vectors (with varying numbers of bits) when new queries are submitted or running queries have finished. However, in practice, dimension' sizes can be large. As a consequence, it may require external hash-joins and therefore resulting in slower performance and unpredictable query execution times.

Our proposal, SPIN, shares some characteristics with those solutions, namely the data sharing, the data pipeline processing, and processing and sharing data scans in a circular loop. While SharedDB uses standard query processing techniques such as index nested-loops, hashing and sorting for any kind of operator of the relational algebra (e.g., joins, grouping, ranking, and sorting), CJoin and DataPath are limited to sharing the join computation and to the cases in which the particular CJoin and DataPath join methods show good performance.

We tackle the dimension size problem using a different approach, which has small memory requirements and can effectively be deployed into parallel shared nothing architectures composed of heterogeneous processing nodes. Our proposal, SPIN is conceptually related to CJoin, and QPipe in what concerns the continuous scanning of fact data, but it uses a simpler approach with minimum memory requirements and does not have the limitations of such approaches. SPIN uses a de-normalized model, as proposed in [8], as a way to avoid joining costs, at the expense of additional storage costs, and to attain massive parallelization [9] with balanced data distribution, scalable performance and predictable query execution times. The CJOIN logic for small in-memory relations and the dynamic scheduling of cooperative scans can be integrated in SPIN.

3   The SPIN Processing Model

SPIN provides workload scale-out by combining the costly IO requests from all the queries into a common data flow (aka pipeline) that is filled by a sequential continuous scan executed in a circular loop. Query execution them proceeds by consuming the relevant data for each query as it flows along the pipeline. This presents a huge potential for data and processing sharing. SPIN uses the de-normalized data model (ONE) proposed in [8, 9], this means that the star schema is physically organized as a single de-normalized relation (O_d).

It views the ONE relation (O_d) logically as a circular relation, i.e. a relation that is constantly scanned in a circular fashion (when the end is reached, it continues scanning from the beginning). The relation is divided into a set of logical fragments (or chunks), with the chunk size adjusted to storage characteristics. The circular loop is continuously spinning, sequentially reading data chunks, while there are queries running. Data is read from storage and shared to all running concurrent queries, as illustrated in Fig. 2.

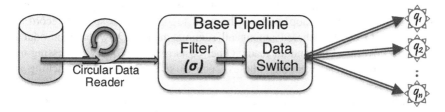

Fig. 2. SPIN Data processing model

A Data Reader sequentially reads chunks of relation O_d, and continuously fills a data pipeline. Tuples when entering the pipeline, have to go through a fast and simple selection operator to early discard large subsets of tuples not required by the currently registered queries (early selection). Only tuples relevant for at least one the queries flow through the pipeline to a Data Switch (DS), which diverts tuples to each of the running queries, building a dedicated logical branch for each query. For performance reasons, some fast early selection predicates are incorporated within the Data Reader.

Since O_d is de-normalized, no costly join tasks need to be processed, only query operations. Any query q, when submitted, starts consuming and processing the tuples that are placed along the data pipeline.

3.1   Query Registering and Processing

A Query Handler handles the query (de)registering. Any query q registers as a consumer of the data in the pipeline, following a publish-subscribe model. Each query q, has to process every tuple from relation O_d, at most once, although data query processing does not need to start at record 0.

For each running query, the Query Handler maintains an indicator of the first logical tuple (position in the circular loop) consumed by the query. The position of this logical first row is fundamental to determine when the end of the query is reached. When that occurs, then the query q has considered all tuples for execution, therefore finishes its execution and sends the results back to the client. Figure 3 illustrates the data reading circular process, depicting the position of logical first row of queries q_1, q_2 and q_3.

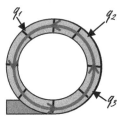

Fig. 3. SPIN sequential data reading loop

Since each query starts consuming tuples from the current position (logical beginning) in the circular loop, without the need to start from a specific record, the reading cost (IO cost) is shared among all running queries Q_r without introducing additional IO overhead or random reads. Other costs related to query processing may be shared at subsequent executing phases, such as selection, logical branching and pipeline processing.

When submitted, a query is analyzed and decomposed into a sequentially-organized set of predicates, computations and aggregations tasks. This decomposition into tasks allows SPIN to determine and update the set of early selection predicates that are placed at the base data pipeline. The remaining query tasks are mapped into SPIN operators for execution. To ensure a fast early selection phase, complex (costly) query predicates are placed at later stages.

3.2   SPIN Operators and Data Processing Pipelines

SPIN follows a flow oriented processing model where each query is decomposed into tasks, which are later mapped to operators placed along a query-specific processing pipeline. A processing pipeline is a collection of sequentially-organized operators that transform tuples as they flow along the pipeline (illustrated in Fig. 4).

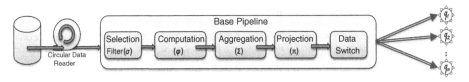

Fig. 4. SPIN Data pipeline processing model

SPIN include the following base operators:

- **Selection Operators - σ** (*conditions*) – apply predicate clauses to filter incoming tuples, trashing those that do not meet the predicate clauses. Each Selection Operation maps a query predicate (default) or a set of related query predicates. The selection operators are typically placed at the beginning of the processing pipeline, to filter tuples that pass-through the pipeline before entering into subsequent processing operators. Selection Operators are placed in a sequential ordered fashion according to their selectivity and evaluation costs, with more restrictive and fastest placed at early stages.

- **Projection operators - π** (*attributes*) – restrict to a subset of tuple attributes that flows throw the pipeline. A projection and a selection operator can be combined into a single step operator.

- **Computation Operators - φ** (*computation expression*) - perform tuple level data transformations, including arithmetic (e.g. $\varphi(a + b)$) and string manipulation (e.g. substring). A dedicated Computational Operator is built for each specific transformation. A Computation Operator maps a tuple-level arithmetic, function or operation expressed in any of the query clauses (e.g. the arithmetic expression QUANTITY * PRICE). The mapping into φ is particularly relevant for complex computations that appear in several query clauses. Query predicates that include computations (e.g. QUANTITY * PRICE > 1000) may be mapped into a σ, preceded by a φ that performs the computation. The goal is to build fast selection operation with simple and fast evaluation predicates.

- **Data Switches - DS** (*attribute conditional switching*) –forward incoming data tuples into a set of data outputs, called **logical data branches (B)**. Tuples are forwarded to all branches b, $b \in B$, or forwarded according to each branch conditions. For each branch, a tuple is only forwarded if it matches the branch's conditions b_p (if exists). Tuples not matching any branch predicates are trashed.

- **Aggregation operators - Σ** (*grouping attributes*; *aggregation functions*) – perform group by computations, by grouping tuples according to the *grouping attribute* clause, and processing the *aggregation functions* (e.g. SUM). Aggregation operators output results when all tuples for a given query as been considered for processing.

A **Data Processing Pipeline** (or simply referenced as pipeline) is a set of sequentially organized operators. A pipeline represents a set of common operations that must be performed to incoming tuples. Each query is decomposed into tasks (e.g. filtering, aggregation) that are mapped into a set of sequential operators and placed a along a query-specific pipeline. Query processing only starts after the pipeline is built and a logical branch is registered as a consumer of the base data pipeline.

3.3 Workload Processing Tree (WT) and Logical Data Paths

The amount of operators (e.g. selection and aggregation operators) and query-specific pipelines (one for each query) increases with the query load, and can rapidly exhaust memory and processing resources. With large concurrent query loads, queries may have common query predicates, computations, and aggregations, resulting in multiple similar operators (each doing its own computation) being placed in query-specific pipelines. The circular Data Reader shares the IO reading cost among queries, but SPIN exploits further data processing opportunities.

Sets of queries may share the same query predicates, computations or aggregations. For each query, SPIN splits the query-specific pipeline into an equivalent ordered set of sequentially connected partial pipelines. Each of these partial pipelines, composed with one or set of logically related operators, is connected as a data consumer of its predecessor. For the currently running query load, similar partial pipelines (with the same operators over the same tuples) from different queries are combined into a common pipeline and a data switch is appended to end to share its results. The subsequent connected data pipelines are then connected as logical branches of this common data pipeline, consuming its output. A set of parallel query-specific pipelines with common operators are rearranged in order to push–forward and to orchestrate similar operators into a common processing pipeline. At the end of this pipeline, a *DS* diverts tuples to further processing in subsequent logical branches.

As a result, the initial query-specific processing pipelines of the currently running queries are split, merged and organized into a workload data processing tree (WT). In the end, the initial query-specific data pipeline of each query will be represented by an equivalent logical data path, which passes through logical branches, *DS* and pipelines.

Figure 5 illustrates a SPIN processing layout with two logical branches composed by two query-specific processing pipelines connected to a common processing pipeline.

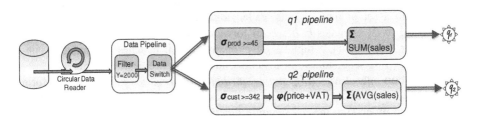

Fig. 5. SPIN Data pipeline processing model

3.4 Building the Workload Processing Tree

The merging of common data processing pipelines and operators is enforced through all the query execution steps, to maximize data and processing sharing and reduce memory and processing usage. For instance, selection operators that are common to running queries are pushed closer to the Data Reader to reduce the data volume within the pipelines and thus increasing the level of data processing sharing. The base pipeline trashes tuples not required by any of its logical branches.

For example, consider that three following aggregation queries are currently running, each with different query predicates:

$$q_1 = SUM\left(\sigma_{p=a}(sales)\right)$$

$$q_2 = SUM\left(\sigma_{y=2000}(sales)\right)$$

$$q_3 = SUM\left(\sigma_{y=2000 \wedge p=a}(sales)\right)$$

For each query, an initial query-specific pipeline is built which must be connected to the base pipeline. Figure 6 depicts the query-specific pipeline, one for each query.

When a new query is submitted, and instead of simply connecting the corresponding query-specific pipeline to the base pipeline, SPIN tries to find intersection points in the current workload processing tree, in order to maximize data and processing sharing with the already running queries. To accomplish this goal, SPIN uses the ***addPath*** algorithm, shown below, to maintain and update the workload processing tree to reflect the changes introduced by the query-specific pipeline of the new submitted query.

Algorithm: addPath
Adding a query-specific pipeline to the Workload Processing Tree (WT)

Input: Q_{path} the query-specific pipeline of query Q
Input: *pipeline* a WT's pipeline, by default the base (root) pipeline
Output: the branch b where the query pipeline is plugged
Data: Workload Processing Tree (*WT*)

1   **if** $Q_{path} = \varnothing$ **then** return;
2   **Let** β be the *pipeline's* branches; i.e. next connected pipelines
3   **if** $\exists\, b \in \beta : b \subset Q_{path}$ **then**
4    $\quad\mid\quad Q_{path} \leftarrow Q_{path} - \{\, b\,\}$;
5    $\quad\mid\quad$ **return** $addPath\,(\,b, Q_{path}\,)$;

6   **if** $\exists\, b \in \beta \wedge p \in Q_{path} : p \cap b \neq \varnothing$ **then**
7    $\quad\mid\quad P_{common} \leftarrow b \cap p;$
8    $\quad\mid\quad P_{rest} \leftarrow b - P_{common};$
9    $\quad\mid\quad Q_{path} \leftarrow Q_{path} - \{\,p\}$;
10   $\quad\mid\quad addPath\,(\,pipeline, Q_{path} + \{P_{common}\})$;
11   $\quad\mid\quad$ **return** $addPath\,(pipeline, Q_{path} + \{P_{rest}\})$;

12   **return** $addBranch\,(\,pipeline, Q_{path})$;

The query-specific pipeline is split into a set of sequentially connected partial-pipelines, called the query data path - Q_{path}. If Q_{path} cannot be fully mapped in WT, i.e. WT does not have an equivalent logical path (T_{path}), then it attempts to match some of Q_{path}'s pipelines. This process is performed to each WT's pipeline, starting from the WT's root pipeline to the leaves, while the current WT's pipeline (b) fully matches a pipeline of Q_{path}. The matching pipelines are removed from Q_{path} and the process is recursively executed for the remaining Q_{path} using b as the new root pipeline.

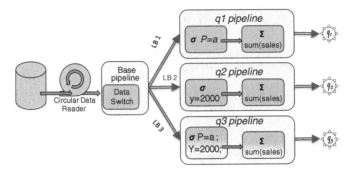

Fig. 6. SPIN deployment of query-specific pipelines

If there's no branch b that fully matches one of Q_{path} pipelines (p), then it tries to find a branch b that partially matches p. When there's a partial match between a WT's pipeline (b) and a Q_{path}'$p \cap b$pipeline (p), i.e. there's a interception between their selection predicate regions, then p is removed from Q_{path} and two distinct paths Q_{path1} and Q_{path2} are created, where $Q_{path1} = Q_{path} + \{p \cap b\}$ and $Q_{path2} = Q_{path} + \{b - (p \cap b)\}$. The algorithm is then applied to each of these paths Q_{path1} and Q_{path2}.

When a pipeline does not matches a Q_{path}'s pipeline, either fully or partially, then remaining Q_{path} is connected to as a new branch of the last matching pipeline. If no matching pipeline exists then Q_{path} is connected to the base pipeline.

The number and placement of DSs, and logical branches, are orchestrated in order to minimize the switching cost DS_{cost}, the number of evaluated predicates, the predicate evaluation costs and the memory requirements for branch management. New logical branches are created and connected to DS when query predicates of processing pipelines queries do not match the predicates of the existing branches.

3.5   Merging and Reusing Intermediate Results

An optimization process transverses all logical branches trying to push-forward operators into the preceding pipelines. The deployment of σ and DS is planned in order to reduce the data volume that flow along the pipelines, and to maximize data and processing sharing.

Afterwards, SPIN applies a merging process that analyses selection predicates and how processing of intermediate results can be reused and shared among processing pipelines, and merged to avoid similar computations of other pipelines. For each logical data path, it follows the path backwards to the source, and at each pipeline **P** of the data path, it determines if exists other logical paths to process, or has already started to be process a subset of the tuples that this pipeline has to process.

When a logical path **LP** exists, then this pipeline is divided in two sequential pipelines (**P1** and **P2**). The latter (**P2**) is connected to **P1** and **LP** and starts consuming their outputs and merging the results. The selection predicates of **P1** are updated to exclude the predicates of the logical path **LP**. This process can result in multiple alternative branching deployments.

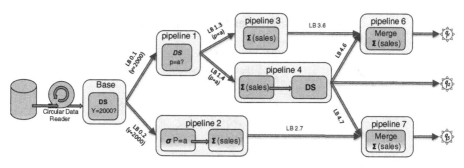

Fig. 7. Merging and reusing intermediate results

To evaluate these alternative deployments, and merging configurations, the merging process uses several data related metrics: n_{tuples} as the number of relation tuples, n_{eval} as the total number of tuples evaluated by σs, and n_{ag} as the total number tuples aggregated by Σs. In the example, the initial deployment, without considering considered merging, the number of evaluated and aggregated tuples are computed, respectively, as

$$n_{eval} = 3 \times n \text{ and } n_{ag} = n_{eval}(\sigma_{p=a\#}) + n_{eval}(\sigma_{y=2000}) + n_{eval}(\sigma_{p=a,y=2000})$$

After the merging process, the number of evaluated tuples n_{eval} which has obtained as a function of the number of running queries Q, is reduced from $Q \times n_{tuples}$ to $\Sigma n_{eval}(p_\sigma)$ with p_σ the selection predicates of p. Figure 7 depicts the final deployment after the merging process.

The number of aggregated tuples n_{ag} is also reduced from $\Sigma n_{eval}(\sigma_q)$ to $n_{eval}(\sigma_{y=2000}) + n_{eval}(\sigma_{p=a,y=2000})$. The total number of evaluated and aggregated tuples are computed, respectively as

$$n_{eval} = n_{eval}(\sigma_{y=2000}) + n_{eval}(\sigma_{y=2000}) + n_{eval}(\sigma_{y\neq2000}) = n_{tuples} + n_{eval}(\sigma_{y=2000})$$
$$n_{ag} = n_{eval}(\sigma_{p=a,y=2000}) + n_{eval}(\sigma_{p=a,y\neq2000}) + n_{eval}(\sigma_{p\neq a,y=2000})$$
$$= n_{eval}(\sigma_{y=2000}) + n_{eval}(\sigma_{p=a,y\neq2000})$$

In this example, we observe that the number of evaluated tuples (n_{eval}) is substantially reduced from 3 times the number of tuples (n_{tuples}) to n_{tuples} plus the number of tuples that satisfy the predicate $(\sigma_{y=2000})$. More than 1/3 of the tuples, depending of the selectivity of $\sigma_{y=2000}$, aren't evaluated. This reduction is even greater as the number of concurrent queries increases and as the overlapping of query predicates increases. Figure 8 illustrates this behavior, using the setup described in Sect. 5.

The results show, as the number of concurrent queries increases, a significant reduction in the number of evaluated tuples by SPIN, while observing an almost linear increase in the number of evaluated tuples of the common query-at-time processing model of most database systems.

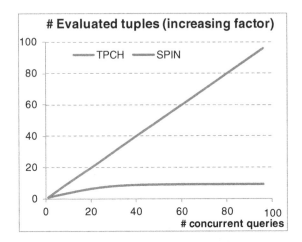

Fig. 8. Number of times tuples are evaluated

3.6   Query Handling and Workload Processing Tree Reorganization

The workload processing tree (*WT*) is continuously reorganized as new queries are submitted. New queries that cannot be directly plugged into *WT* are kept as distinct branches connected to the base (root) pipeline.

When a query finishes its execution, the Query Handler, removes the query-specific pipelines, which are not used in other logical paths, from the workload processing tree. Pipelines used by other logical paths are maintained, only the query-specific outputs are detached. Query-specific logical branches are detached and removed. Then a reorganization process is triggered to update the workload processing tree.

For instance, in the previous example (Fig. 7), when the query q3 finishes its execution, the query-specific pipeline 7 is detached from both pipeline 2 and 4 before being removed.

When a pipeline is removed, it triggers a reorganization process to update the workload processing tree. The goal is to determine if exists *WT* pipelines that are producing output results that aren't consumed (i.e. without connected outputs), and therefore can be removed from the *WT* (Fig. 9).

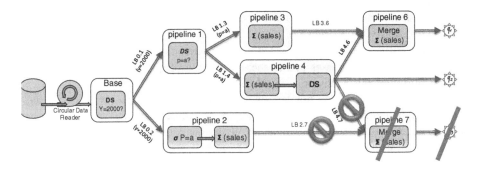

Fig. 9. WT reorganization with group removal – step 1

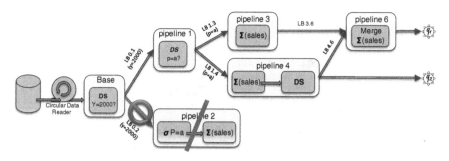

Fig. 10. *WT* reorganization with group removal - step 2

In pipeline 2, the aggregation operator (e.g. SUM (sales) of $\sigma_{y=2000}$ and $\sigma_{p=a}$) that is not used by any of the running queries, can be excluded from query processing. As a result, the related processing branch (pipeline 2 in the example) can also be removed from the workload processing tree (reducing memory usage and data processing). The selection predicates can be pushed to preceding pipelines, in order to reduce the amount of data to be processed and that flows through the pipelines. Pushing selection predicates to preceding pipelines may cause the removal of additional branches, with the associated benefits (Fig. 10).

Since pipeline 2 is removed, the data switch in base pipeline is replaced by a selection predicate $\sigma_{y=2000}$ and the data switch and the logical branch of pipeline 1 is pushed to the base pipeline. Then pipeline 1 is also removed. Operators in use by other running queries are updated to reflect the removal of query-specific clauses. Figure 11 depicts the final *WT* layout.

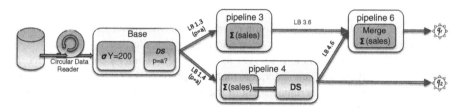

Fig. 11. *WT* reorganization with group removal- final layout

After the *WT* reorganization is completed, a data branching optimization process is triggered to determine if other logical branching deployment can deliver improved performance.

4 SPIN Prototype

We have built a SPIN prototype implemented in Java to evaluate its performance and scalability capabilities. This section presents details of the SPIN prototype, which implements the mechanism discussed above. The prototype was built as a set of flexible

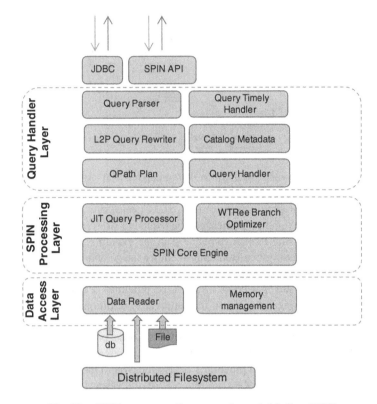

Fig. 12. SPIN prototype diagram - release 1.6.3 (June2013)

and extensible modules, organized in three main layers (illustrated in Fig. 12): a data access layer, a SPIN core processing layer and a query handler layer.

The SPIN prototype offers a SPIN API for querying and interacting with SPIN. It also offers a JDBC compliant interface that interprets SQL-92 SELECT queries, allowing SPIN to be used as a replacement of the existing DW infrastructure or be used as query accelerator.

The illustrated prototype in Fig. 12 relates to the release 1.6.3 (June2013), which has about 32klocs and 153 java classes and do not include several of SPIN optimization mechanisms, such as: data columnar storage organization, compression, partial de-normalized relations with in-memory dimensions, massive data sharding, data loading and snapshot isolation.

4.1   The Data Access Layer

The data access layer implements all the functionalities related to gathering the data from storage to the base pipeline for processing. Data is gathered by its main module, the data reader, which can be extended to handle distinct storage locations and physical data organizations. The default data reader accesses tuples physically stored in a row-wise format in a full de-normalized relation as proposed in [8]. We have implemented several data readers, including sequential and mapped memory file using row-wise tuple organization, in-memory object-oriented tuple organization, compressed row-wise, and SQL based data gathering from common databases using a JDBC driver. The later data reader allows SPIN to act as a middle-layer query accelerator of existing DW infrastructures. This layer also manages a buffering (caching) memory to speedup the data access time, and the processing of frequent accessed data.

4.2 The Query Handler Layer

The query handler layer is responsible for handling query requests, for parsing and performing a syntactical and object validation against the information in the metadata catalog. It also contains a module that extracts, if exists, and validates the timely related clauses, to check if they can be satisfied under the current query load. As discussed above, SPIN by default uses a denormalized ONE storage organization to avoid joining relations, particularly large dimensions that cannot fit entirely in memory, and therefore providing predictable execution times, but a distinct data reader can be implemented to use other storage organization. SPIN, in order to be a used as a middle-layer and/or a transparent replacement of the existing DW infrastructure, it maintains a logical star-schema representation of the DW model, regardless of the physical storage organization.

A Logical-to-Physical translator module rewrites the star-schema queries according to the SPIN's physical storage model. It provides a JDBC interface to allow seamless integration with users and applications and the submitted SQL queries are syntactically analyzed according to the logical star-schema view. As the physical schema may diverge from the logical star-schema view, queries are rewritten according to the physical schema representation, into a set of simpler processing tasks with more predictable execution time. A query Q, syntactically valid according to the logical schema, is translated (rewritten) into query Q_t (or a set of sub-queries) according to the internal physical organization. Appendix A shows some TPC-H queries and their equivalent translation made by SPIN before being processed.

Afterwards, a query planner and optimizer builds a Q_{path} execution plan for each rewritten query. A query handler manages the execution and completion of the query Q_{path} plan execution and triggers the WT reorganization when the query completes (reaches the first logical row).

4.3 The SPIN Processing Layer

The SPIN processing layer handles the query execution maximizing the data and processing sharing among queries. It implements the algorithms discussed in Sect. 3.1 to plug the specific query pipeline (Q_{path}) to the currently running workload processing tree (*WT*). Then a just-in-time (JIT) query processor uses dynamic coding, to implement all the specific operators, pipelines and data branches required to process the running queries. Afterward the query execution can be started by the SPIN core engine.

A WTree branch optimizer is continuously monitoring the execution, the addition and conclusion of running queries. A workload processing tree reorganization is triggered whenever a different deployment can provide higher data and processing sharing and yield better performance.

5 Evaluation

This section discusses experimental evaluation results obtained by the SPIN prototype described above (release 1.6.3), which implements the mechanisms discussed in the paper. We used the default data reader that reads tuples in a row-wise format physically stored in a full de-normalized relation as proposed in [8], with disabled optimization features (e.g. compression, materialized views, automatic in-memory bit-selections).

The setup is based on an Intel i5 processor, with 8 GB of RAM and a RAID0 storage system composed with 3 SATAIII disks with 2 Terabytes each, running a default Linux Server distribution. An additional server, connected through a gigabit-Ethernet switch, was used for submitting a varying concurrent query load. The server runs a default installation of PostgreSQL 9.0 [10], with shared data buffers set to 2 GB. We also evaluated SPIN with DbmsX (a well known commercial RDBMS) and obtained similar results.

We used the TPC-H benchmark with scale factors 1 (SF1) and 10 (SF10) and build two distinct setups to evaluate SPIN. The TPCH setup which was populated with the TPC-H data generator tool (DBGEN) available at [11] and the SPIN setup which access a ONE relation populated with a modified version that generates the de-normalized data as single flat file.

To evaluate the impact of the concurrent query load in performance and data volume, we used a query load composed by several variants of the query Q5 with different selectivity (different date ranges: 1 year, x days, ...) and aggregation groups (e.g. n_name or n_name, c_mktsegment, ...). We also used a more complex workload composed with variants of the queries Q1, Q3 and Q5 with distinct query predicates to evaluate the influence of the workload query pattern in the average execution time. Appendix A shows an example of the original submitted queries and also the equivalent queries after translated by the L2P Query Rewriter before being processed by SPIN.

The query load consisted in a total of 1000 queries chosen randomly among these variants, each with a random number of filter predicates and random values in the filter predicate (e.g. X days). The queries were submitted concurrently by a varying number of simultaneous clients. The results depicted below were obtained as the average of 30 runs.

5.1 Influence of Number of Queries in Query Performance

Query execution time of common RDBMS that follow a query-at-time execution model, is highly influenced by the number of queries that are concurrently being executed. In this setup, we evaluate how the number of concurrent queries influences

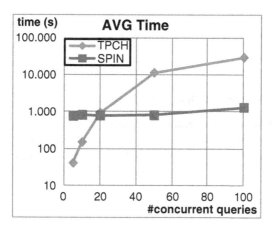

Fig. 13. Average execution time for varying query loads (lower is better)

the average execution of SPIN and TPCH. Figure 13 depicts the average execution time for a scale factor of SF = 10.

We observe that with low concurrent query loads (less than 20 concurrent queries), the TPCH setup yields better average execution times. However as the number of concurrent queries increases, the TPCH setup exhibits significantly higher average execution times because more queries are competing for resources. On the other hand, the average execution time with the SPIN setup remains almost constant. There's a slight increase at higher concurrent query loads due to the pipeline management overheads and the cost of processing the query-specific pipelines that cannot be combined with other query pipelines.

Figure 14 depicts the impact of submitting additional queries in the average execution time of the currently running queries. The results show that at higher query loads, SPIN introduces low overheads per query (below 1 %) in the average execution time. The overhead is higher at low query loads (less than 10 concurrent queries) because the running queries exhibit less opportunities for data and processing sharing.

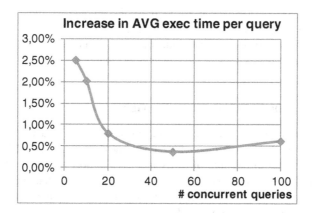

Fig. 14. Overhead per query in the average execution time (lower is better)

The results also reveal that, when a query is submitted using SPIN, there's a high degree of confidence regarding how long it will take to deliver the result, regardless of the currently running query load.

5.2 Influence of Number of Queries in Throughput

Since the average execution time is not significantly influenced by the concurrent query load, SPIN yields almost linear throughput. The probability that two or more queries may share overlapping selection predicates and processing operators, increases with the number of concurrently running queries and therefore yielding improved throughput (Fig. 15).

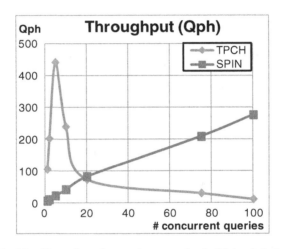

Fig. 15. Throughput for varying query loads (higher is better)

We observe that with low concurrent query loads, the TPCH setup yields higher throughputs than SPIN. However as we increase the number of queries running concurrent this behavior changes drastically. SPIN does not deliver a linear throughput due to the pipeline management overheads and the cost of processing the query-specific pipelines that cannot be combined with other query pipelines.

5.3 Influence of the Data Volume in Throughput

Throughput is influenced by the query load, but also by the data volume. Figure 16 depicts the throughput for two distinct data volumes: SF1 (a) and SF10 (b). In the figure, we observe that throughput of SPIN increases almost linearly with the number of concurrent running queries, as more data and processing is shared among queries. With SF1, TPCH yields significantly higher throughput since all data and processing is done almost exclusively in memory. However as the number of concurrent queries increase we observe a significant drop in throughput as the running queries exhaust the available memory. SPIN, does not have these memory issues and can be massively partitioned among low-end commodity servers.

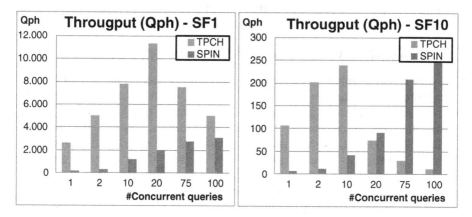

Fig. 16. Throughput for varying query loads with (a) SF = 1 and (b) SF = 10 (higher is better)

This effect can be observed in Fig. 17, which compares the throughput ratio (Qph (SF1)/Qph(SF10)). As the data volume increases by a factor of 10, from SF = 1 to SF = 10, we observe that at low query workloads (less than 10 queries running concurrently), the throughput drops by a factor of around 30, mainly because with SF = 1, TPC-H is remains almost entirely in memory, while with SF = 10 there's more IO operations. As the query workload increases loads, TPCH shows an increasingly larger drop in throughput, since more IO operations are required to process queries. With 100 concurrent queries, an increase in data volume by factor of 10 results in a

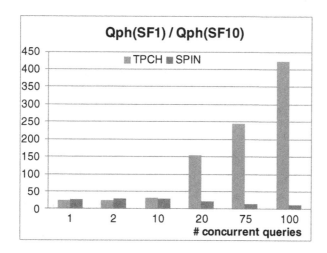

Fig. 17. Impact in throughput of a 10x increase in data volume (lower is better)

drop in throughput by a factor greater than 400. On the other hand, the throughput of SPIN drops almost proportionally to the data volume increase factor.

For a higher data volume, the TPCH setup will experience a higher drop in throughput since it will require more IO operations to process the joins, while SPIN will yield a drop in throughput, proportional to the data volume increase.

5.4   Influence of the Workload Query Pattern in Query Performance

Since data is sequentially read and shared by the running queries, the IO cost remains constant regardless of the query pattern. But in Fig. 13 we observed a slight increase in the average execution time as the number of concurrent queries increases, mainly because the workload processing tree becomes wider, as more queries as being simultaneous processed, and also due to the complexity of the evaluation predicates. The above experiments were carried out using variants of the query Q5 but with different predicate clauses, selectivity and aggregation groups.

We now compare these average execution times with more complex query loads, to evaluate the influence of the query workload pattern in the average execution time. Figure 18 depicts SPIN results for the SF = 10 with three distinct query workloads: Q1 and Q5 are query workloads exclusively composed by variants of query Q1 and Q5, respectively; Q135 is a workload composed by a set of variants of the queries Q1, Q3 and Q5 (Appendix A), that where randomly chosen for execution.

The results show that the query workload has minimum impact in the average execution time, for all the considered number of concurrent queries. Although some queries have more complex computations, there are no significant changes in the average execution time because for such concurrent query workloads the CPU hasn't

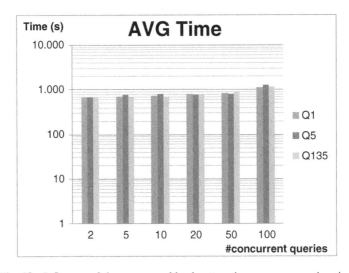

Fig. 18. Influence of the query workload pattern in average execution time

the bottleneck, but the IO performance. Therefore, and because it uses the scalable and predictable ONE model, SPIN performance can be significantly boosted if a parallel infrastructure is used.

6 Conclusions

We presented the mechanisms of SPIN, a data and processing sharing model that deliver predictable execution times to star-join queries even in the presence of large concurrent workloads, without the memory and scalability limitations of existing approaches. We described the mechanisms used by SPIN to embed data and queries in a shared workload query processing tree and how SPIN dynamically reorganizes the processing tree. We described the implementation details of the SPIN prototype used in experimental evaluation and used the TPC-H benchmark to assess its ability to provide scalable performance and predictable execution times, even in presence of large concurrent query loads.

Currently, we are undergoing a more exhaustive performance analysis, using a larger subset of the TPC-H queries, and we are extending the SPIN processing model to deliver assured time guarantees with large parallel heterogeneous deployments, with massive data sharding.

Appendix A

Query Q1 (Original):
```
SELECT l_returnflag, l_linestatus,sum(l_quantity) as sum_qty,
    sum(l_extendedprice) as sum_base_price,
    sum(l_extendedprice*(1-l_discount)) as sum_disc_price,
    sum(l_extendedprice*(1-l_discount)*(1+l_tax)) as sum_c,
    avg(l_quantity) as avg_qty,
    avg(l_extendedprice) as avg_price,
    avg(l_discount) as avg_disc,
    count(*) as count_order
FROM lineitem
WHERE l_shipdate <= date '1998-12-01' - interval '30' day
GROUP BY l_returnflag,    l_linestatus
ORDER BY l_returnflag, l_linestatus;
```

Query Q1 (Translated):
```
SELECT l_returnflag, l_linestatus,sum(l_quantity) as sum_qty,
    sum(l_extendedprice) as sum_base_price,
    sum(l_extendedprice*(1-l_discount)) as sum_disc_price,
    sum(l_extendedprice*(1-l_discount)*(1+l_tax)) as sum_c,
    avg(l_quantity) as avg_qty,
    avg(l_extendedprice) as avg_price,
    avg(l_discount) as avg_disc,
    count(*) as count_order
```

```
FROM spin_tpch
WHERE l_shipdate <= date '1998-12-01' - interval '30' day
GROUP BY l_returnflag,    l_linestatus
ORDER BY l_returnflag, l_linestatus;
```

Query Q3 (Original):

```
SELECT l_orderkey, o_orderdate, o_shippriority
   sum(l_extendedprice * (1 - l_discount)) as revenue,
   FROM customer, orders, lineitem
WHERE  c_mktsegment = 'BUILDING'
   and c_custkey = o_custkey
   and l_orderkey = o_orderkey
   and o_orderdate < date '1995-03-15'
   and l_shipdate > date '1995-03-15'
GROUP BY l_orderkey, o_orderdate, o_shippriority
ORDER BY revenue desc, o_orderdate;
```

Query Q3 (Translated):

```
SELECT l_orderkey, o_orderdate, o_shippriority
   sum(l_extendedprice * (1 - l_discount)) as revenue,
FROM   spin_tpch
WHERE  c_mktsegment = 'BUILDING'
   and o_orderdate < date '1995-03-15'
   and l_shipdate > date '1995-03-15'
GROUP BY l_orderkey, o_orderdate, o_shippriority
ORDER BY revenue desc, o_orderdate;
```

Query Q5 (Original):

```
SELECT n_name,
   sum(l_extendedprice * (1 - l_discount)) as revenue
FROM   customer, orders, lineitem, supplier, nation,region
WHERE  c_custkey = o_custkey
   and l_orderkey = o_orderkey
   and l_suppkey = s_suppkey
   and c_nationkey = s_nationkey
   and s_nationkey = n_nationkey
   and n_regionkey = r_regionkey
   and r_name = 'ASIA'
   and o_orderdate >= date '1994-01-01'
   and o_orderdate < date '1994-01-01' + interval '1' year
GROUP BY n_name
ORDER BY revenue desc;
```

Query Q5 (Translated):

```
SELECT n_name,
   sum(l_extendedprice * (1 - l_discount)) as revenue
FROM   spin_tpch
WHERE  r_name = 'ASIA'
   and o_orderdate >= date '1994-01-01'
   and o_orderdate < date '1994-01-01' + interval '1' year
GROUP BY n_name
ORDER BY revenue desc;
```

References

1. Candea, G., Polyzotis, N., Vingralek, R.: A scalable, predictable join operator for highly concurrent data warehouses. Proc. VLDB Endow. **2**, 277–288 (2009)
2. Candea, G., Polyzotis, N., Vingralek, R.: Predictable performance and high query concurrency for data analytics. VLDB J. **20**(2), 227–248 (2011)
3. Zukowski, M., Héman, S., Nes, N., Boncz, P.: Cooperative scans: dynamic bandwidth sharing in a DBMS. In: Proceedings of the 33rd International Conference on Very Large Data Bases, Vienna, Austria, pp. 723–734 (2007)
4. Harizopoulos, S., Shkapenyuk, V., Ailamaki, A.: QPipe: a simultaneously pipelined relational query engine. In: Proceedings of the 2005 ACM SIGMOD International Conference on Management of Data, pp. 383–394 (2005)
5. Unterbrunner, P., Giannikis, G., Alonso, G., Fauser, D., Kossmann, D.: Predictable performance for unpredictable workloads. Proc. VLDB Endow. **2**, 706–717 (2009)
6. Arumugam, S., Dobra, A., Jermaine, C.M., Pansare, N., Perez, L.: The DataPath system: a data-centric analytic processing engine for large data warehouses. In: Proceedings of the 2010 International Conference on Management of Data, pp. 519–530 (2010)
7. Giannikis, G., Alonso, G., Kossmann, D.: SharedDB: killing one thousand queries with one stone. Proc. VLDB Endow. **5**(6), 526–537 (2012)
8. Costa, J.P., Cecílio, J., Martins, P., Furtado, P.: ONE: a predictable and scalable DW model. In: Cuzzocrea, A., Dayal, U. (eds.) DaWaK 2011. LNCS, vol. 6862, pp. 1–13. Springer, Heidelberg (2011)
9. Costa, J.P., Martins, P., Cecílio, J., Furtado, P.: A predictable storage model for scalable parallel DW. In: Fifteenth International Database Engineering and Applications Symposium (IDEAS 2011), Lisbon, Portugal (2011)
10. PostgreSQL. http://www.postgresql.org/
11. TPC-H Decision Support Benchmark. http://www.tpc.org/tpch/

An Uncoupled Data Process and Transfer Model for MapReduce

Li Zha[1], Jie Zhang[1,2], Wei Liu[1,2(✉)], and Jian Lin[1,2]

[1] Institute of Computing Technology, Chinese Academy of Sciences, Beijing, China
char@ict.ac.cn, {zhangjie,liuwei,linjian}@software.ict.ac.cn
[2] University of the Chinese Academy of Sciences, Beijing, China

Abstract. In the original MapReduce model, reduce tasks need to fetch output data of map tasks in the manner of "pull". However, reduce tasks which are occupying reduce slots cannot start executing until all the corresponding map tasks are completed. It forms the dependence between map and reduce tasks, which is called the coupled relationship in this paper. The coupled relationship leads to two problems: reduce slot hoarding and underutilized network bandwidth. Meanwhile, storing the result data is costly especially when the system has replications, which leads to the inefficient storage problem. We propose an uncoupled data process and transfer model in order to address these problems. Four core techniques, including weighted mapping, data pushing, partial data backup, and data compression are introduced and applied in Apache Hadoop, the mainstream open-source implementation of MapReduce model. This work has been practiced in Baidu, the biggest search engine company in China. A real-world application for web data processing shows that our model can improve the system throughput by 29.5 %, reduce the total wall time by 22.8 %, provide a weighted wall time acceleration of 26.3 %, and reduce the result data stored in disk by 70 %. What's more, the implementation of this model is transparent to users and compatible with the original Hadoop.

Keywords: MapReduce · Data transfer · Uncoupled model · Compression

1 Introduction

With the arrival of "big data" era, the original computing and storing systems face great challenges. Platforms which can process and store large data are receiving more and more attention, such as MapReduce [12], Dryad [16], Sector/Sphere [15], and BigTable [8]. The MapReduce programming model proposed by Google has become the mainstream data-centric platform for large data processing because of its scalability and simplicity. Apache Hadoop [1], an open-source implementation of MapReduce, is widely used.

The MapReduce model is a software architecture for parallel computing on large data sets with commercial hardware. A job is divided into map tasks and reduce tasks. Map tasks are responsible for reading the source data, resolving

© Springer-Verlag Berlin Heidelberg 2015
A. Hameurlain et al. (Eds.): TLDKS XVII, LNCS 8970, pp. 24–44, 2015.
DOI: 10.1007/978-3-662-46335-2_2

the data into key-values, and writing the intermediate result into local disk. Reduce tasks read the intermediate result written by corresponding map tasks, resolve the key-values of the same key and write the final result into file system. In the MapReduce architecture, one node works as the master where a Job-Tracker runs. The JobTracker is responsible for monitoring and managing map and reduce tasks. The other nodes work as slaves where TaskTrackers run. The TaskTrackers are responsible for executing map and reduce tasks. When a job is submitted, the related input data is divided into several splits. The JobTracker will pick up idle TaskTrackers to perform map tasks on the splits, and then perform reduce tasks on the intermediate output of map tasks. The final result will usually be a set of key-value pairs stored in a distributed file system like HDFS [22] and GFS [13].

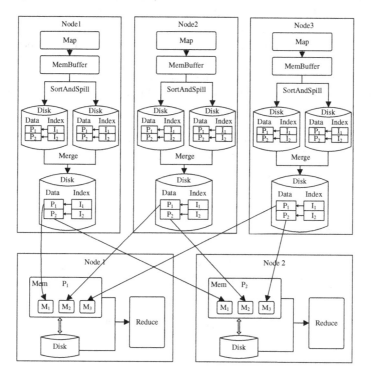

Fig. 1. The data flow in original MapReduce

In the original MapReduce model, data is transferred between map tasks and reduce tasks in the way that can be described as "pull". The reduce task is divided into three phases: shuffle, sort and reduce. When some of the map tasks are completed, the corresponding reduce tasks can fetch data in the shuffle phase. However, the reduce phase of the reduce task will not start until all the map tasks finish. Thus, the reduce tasks will occupy the assigned slots all the time to wait for the completion of all map tasks, which is called the coupled relationship. Due to the coupled relationship between map and reduce tasks, it results in two problems: reduce slot hoarding [26] and underutilized network bandwidth. Meanwhile, as a fault-tolerant computing model, MapReduce introduces replicated distributed storage, which leads to inefficient storage in the data pulling scene (Fig. 1).

1.1 Reduce Slot Hoarding Problem

In the MapReduce model, it usually begins to schedule reduce tasks when a certain amount of map tasks are completed. If a big job is submitted, reduce tasks of the job will occupy all the assigned slots until all the map tasks finish. Consequently, when another job is submitted at this moment, it will not get corresponding reduce slots even after all the map tasks finish. Therefore, the later one will starve until the big job is completed. This is called the reduce slot hoarding problem, which will seriously reduce the execution efficiency of jobs, especially for small ones.

Figure 2 shows an example of reduce slot hoarding problem. Job1 and job2 are submitted at the same time while job3 is submitted a little later. The reduce task of job1 start when a certain amount of the map task finished, but no reduce tasks can be finished before all map tasks finished, so the reduce slots will be occupied by job1 for a long time. In this case, the reduce task of job3 can't execute when its map tasks are finished since job1 occupies all the reduce slots. Job3 will starve until job1 releases the reduce slots.

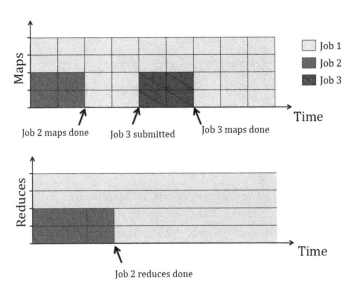

Fig. 2. Reduce slot hoarding problem

One solution to this problem is to delay the reduce tasks. In [26], the authors put forward a solution that starts reduce tasks after the completion of map tasks. However, their tests show that it will decrease the whole throughput. We consider that if reduce tasks start early, partially overlapped with map tasks, they will get a part of the data from map tasks, which can save the time of data transfer and the total completion time.

1.2 Underutilized Network Bandwidth Problem

For a MapReduce job, the network load will mainly concentrate in reduce tasks. In the reduce task, both shuffle and reduce phase need large network bandwidth since shuffle phase should pull data from other nodes, and reduce phase should write the result data into distributed file system. However, map tasks almost don't need large network bandwidth, because MapReduce has done some data locality optimization [7,20,21] when scheduling to make sure map tasks execute on the node where the needed data is stored on. So a map task can mostly read data from local disks.

There are some capabilities to balance network load and optimize network bandwidth in the original version of MapReduce, such as scheduling reduce tasks in advance. When the completion of map tasks reaches a certain ratio (default 5 %), the reduce task will start so that they can run in parallel with maps. However, it can only alleviate this problem rather than resolve it. When some of the early-scheduled reduce tasks, whose desired intermediate data is ready, are getting partitions through shuffle, they cannot use much bandwidth in the map task. Conversely, they will occupy the reduce slots all the time. Besides, limited by the total slots, not all the reduce tasks can be scheduled. Therefore, these reduce tasks cannot work with map tasks simultaneously, and the network bandwidth will still be underutilized.

1.3 Inefficient Storage Problem

The MapReduce model is designed to run on big data sets with commercial hardware. In original version of MapReduce, the result is written into distributed file system directly. It will need many disks to store the data when the result is huge, which is costly especially when the system has one or more replications. This is the inefficient storage problem.

Compressing the result of MapReduce is a good way to reduce the data that is needed to be stored on the disk. Many compression modules have been proposed to address the inefficient storage problem. Apache Hadoop, an open-source implementation of MapReduce, can also support compressing the data now. Hadoop allows users to choose one compression algorithm. Once chosen, all the data will be compressed using this algorithm. BlobSeer [18,19], a typical compression module that has achieved a great success on distributed file system, is a transparent compression module using prediction to determine whether to compress the data or not. First, BlobSeer samples part of the data to predict the compression ratio of total data block, then they judge if compressing the data is beneficial to the system. This module only compresses the data when they think that compression is beneficial to the system. BlobSeer can save about 40 % space comparing with storing the data directly. However, these two modules don't take CPU rate and memory usage into consideration. As we all know that compression can cost too much CPU and memory resource, these modules may have a bad effect on other jobs if CPU is overload or memory is exhausted.

In this paper, we propose an uncoupled MapReduce model to address the above three problems, which can improve the system throughput and overall resource utilization. The rest of this paper is organized as follows. In Sect. 2, we present the uncoupled MapReduce model. Section 3 describes the architecture and implementation of this model. Section 4 offers the evaluation about our model and its application effects. Section 5 lists some related work. At last, Sect. 6 concludes the paper.

2   The Uncoupled MapReduce Model

An uncoupled MapReduce model with intermediate data transfer is designed to address the three problems, meanwhile improving the job execution efficiency and system throughput. Figure 3 shows the data flow in this model. The data transfer is completed during the map task in the uncoupled version of MapReduce, instead of during the reduce task as the original version does. It needs to meet three conditions as follows:

- To address the reduce slot hoarding problem thoroughly, reduce tasks should be scheduled after all map tasks are completed. Reduce tasks will not occupy the slots, and they can read the data to process locally once launched, which improves job execution efficiency and system throughput.
- To address the underutilized network bandwidth problem, the data transfer process of maps' result should be completed in the map task. When reduce tasks start, they will read data directly from local disks. Therefore, the network load will not concentrate in the reduce task and the network bandwidth in the map task can be used fully.
- To address the inefficient storage problem, the result data of MapReduce jobs should be compressed before stored in file system. Compression should be done transparently and automatically. Considering compression jobs may cause overload of CPU, a specific hardware is needed to help reducing the workload of CPU.

However, there is a conflict between the fist two conditions. Map tasks need to transfer data to reduce tasks, but reduce tasks do not run until all the map tasks are completed. Therefore, some trade-offs are necessary to find out when and where reduce tasks run. The inefficient storage problem is more independent relatively to the first two conditions. In order to resolve these problems, we introduce the following four techniques.

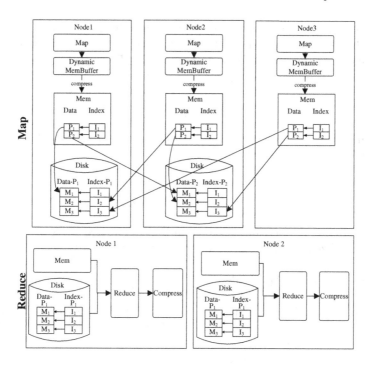

Fig. 3. The data flow in uncoupled MapReduce

– **Weighted mapping** [23]. This technique creates a mapping relationship between reduce tasks and nodes. Through the mapping relationship, the map task can find out the node where the corresponding reduce tasks will run, and they can transfer data to these nodes. In a heterogeneous cluster, the computing capability of each node is not identical. In consideration of this fact, each node has its own weight. The node with higher weight will get more tasks. This technique can guarantee balance and consistency of task assignment.

  • **Balance.** The balance means that the node with greater weight will be assigned more reduce tasks. A linear relationship exists between the number of assigned tasks and the node's weight. We assume that there are n nodes with weight $w_i(1 \leq i \leq n)$ in a cluster. Normalize the weight and get the normalization value w'_i from w_i, as shown in Eq. 1.

$$w'_i = \frac{w_i}{\sum_{i=1}^{n} w_i} \qquad (1)$$

  The variable M stands for the total number of reduce tasks, and M_i stands for the number of reduce tasks assigned to node i. Equation 2 shows the result of task assignment.

$$M_i = w'_i M = \frac{w_i}{\sum_{i=1}^{n} w_i} M \qquad (2)$$

  The two equations above can ensure the balance of task assignment.

- **Consistency.** The consistency means that the fixed mapping relationship between reduce tasks and nodes should be guaranteed and cannot be changed once decided. We assign reduce task $R_j (1 \leq j \leq M)$ with weight w_{R_j}. The relationship is expressed in Eq. 3.

$$w_{R_j} = \frac{j}{M} \qquad (3)$$

Reduce task R_j with weight w_{R_j} will be mapped to node $k(1 \leq k \leq n)$ if they follow the relationship in Eq. 4.

$$\sum_{i=1}^{k-1} w_i' < w_{R_j} \leq \sum_{i=1}^{k} w_i' \qquad (4)$$

The mapping relationship is shown in Fig. 4. The two kinds of weights will be normalized into the same range, and then we can establish the mapping relationship between reduce tasks and nodes if they have the same weighs after normalizing. Any module of a MapReduce application can inquire the relationship through Eq. 4.

Weighted mapping technique has no conflict with the scheduler in original MapReduce module. The original MapReduce module focuses on scheduling map tasks and reduce tasks between different MapReduce jobs while weighted mapping technique focus on the nodes and reduce tasks in one MapReduce job.

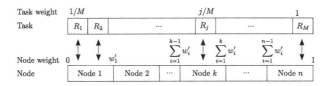

Fig. 4. The mapping relationship between reduce tasks and nodes

- **Data pushing.** In the uncoupled version of MapReduce, map tasks implement data transfer using data pushing. As shown in Fig. 3, map tasks will put intermediate data into dynamic buffers and partition them. Then they push data to the reduce tasks in corresponding mapped node from partition 1 to n in order, which is infeasible in the original version because where reduce tasks are responsible for getting data from map tasks. This idea is partially inspired by the pipelined MapReduce [11]. In the uncoupled version, a server is setup in each node which is responsible for receiving data from map tasks as shown in Fig. 5. When map tasks are generating intermediate output data, they will work as clients and push data to the servers. So there is no need to start reduce tasks before map tasks finish.
- **Partial data backup.** Each node has a server for receiving data from map tasks. If some servers go wrong, the data pushed by map tasks will be lost. It's costly to re-execute map tasks to get the lost data. The original version of

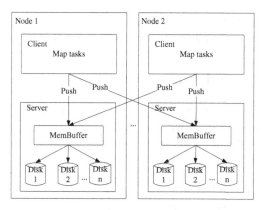

Fig. 5. The diagram of data pushing

MapReduce has the fault-tolerance module to avoid re-executing map tasks. However, uncoupled MapReduce changes the data transfer module leading to the useless of the original fault-tolerance module. The partial data backup technique in uncoupled MapReduce can resolve this problem. When the map tasks are pushing data, the data will also be backed up on local disks. When some servers go wrong, there is no need to re-execute the completed map tasks, because the data can be recovered by reduce tasks. A backup server is setup on each node which is responsible for managing the backup data. Reduce tasks will pull their own backup data by requesting each backup server. "Partial" here means if a partition is pushed from a map task to its own node, it will not be backed up. What's more, it is compatible with the original fault tolerance mechanism in MapReduce.

– **Data compressing.** In the uncoupled version of MapReduce, the result data of MapReduce jobs is stored into file system after compression which can reduce the data amount dramatically. CPU may be overload since the compression jobs need a lot of computing. In this case, other executing jobs may be seriously affected due to CPU resources exhausted. In data compressing technique, we choose the most appropriate compression algorithm to do compression jobs according to the CPU workload, the data property and so on. The technique will transfer compression jobs to a specific hardware if we think the workload of CPU is too heavy. All the job concerning compression include algorithm selecting and compressing are done transparently, and automatically to the user.

This model uncouples the dependency relationship between maps and reduces in MapReduce, replaces the shuffle phase in the original version and compresses the result data of MapReduce jobs. Through the four techniques mentioned above, it can make sure that all the reduce tasks can read data from their local disks when the map tasks are completed and all result data is stored after compressed. Therefore, reduce tasks will not occupy the reduce slots to wait for the completion of map tasks. It can also satisfy the needs of slots from other jobs.

This model can make full use of the network bandwidth in map and reduce tasks, balance the network load and improve the efficiency of storage.

3   Architecture and Implementation

Considering the original architecture of MapReduce and the requirements of four techniques mentioned above, the architecture of uncoupled MapReduce is designed. It includes four kinds of modules: master control module, data transfer module, fault tolerance module, and data compress module. We have implemented the architecture in Hadoop, and integrated different modules in the master and slave nodes. The uncoupled MapReduce architecture and its implementation in Hadoop are presented in Fig. 6.

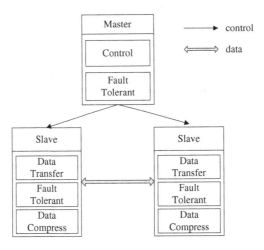

Fig. 6. The uncoupled MapReduce architecture and its implementation in Hadoop

3.1   Master Control Module

The master control module lies in the master node, responsible for monitoring and scheduling tasks, and coordinating other modules. This module should run all through the MapReduce job. Its functionalities are as follows.

- Create the mapping relationship between reduce tasks and nodes to make sure that reduce tasks will be executed on the node where data stored.
- Convey the mapping relationship information to the data transfer module and the fault tolerance module.
- Schedule tasks and make sure that reduce tasks will not start until all the map tasks are completed. When scheduling reduce tasks, make sure all the data is transferred to the specific node.

- Ensure balance and consistency of the mapping relationship, configure the computing capability of every node and make each node getting proper tasks according to its computing capability.
- Coordinate the data transfer module and the fault tolerance module. Normally, this module controls the data transfer module to complete the task of data pushing. If there is something wrong with data pushing, this module will notify the fault tolerance module to recover the missing data.

3.2   Data Transfer Module

The data transfer module lies in all the slave nodes, responsible for processing, storing, and pushing data. The module creates a server responsible for receiving and managing the intermediate data for every slave nodes since the data should be pushed to the node where the corresponding reduce tasks run according to the mapping relationship before reduce tasks begin in uncoupled version of MapReduce. The functionalities of this module are as follows.

- Create a data transfer server in each slave node.
- Get the output data from the map tasks.
- Do some preprocessing on the data.
- Get the mapping relationship.
- Work as a client to push the data of each partition to their corresponding reduce tasks.
- The data transfer server in this module is responsible for receiving and managing data.

3.3   Fault Tolerance Module

The fault tolerance module lies in all the nodes, responsible for processing exceptions caused by system crash, power outage and so on. As our model changes the intermediate data transfer mode, we must make some supplements to the original fault tolerance mechanism. In this module, the partial data backup technique is introduced following specific rules. The functionalities of this module are as follows.

- Backup each partition that map tasks will push to other nodes in the local disk.
- Assign the mapped reduce tasks to other nodes when a node failed.
- Offer the backup data to reduce tasks through backup servers, if the backup data has been made successfully. A reduce task checks whether the node is the mapped one through weighted mapping mechanism. If it is not the mapped one, the reduce task will pull its own backup data by requesting to other backup servers.

The fault tolerance module backups the partitions and the data transfer module pushes the partition to the node where reduce tasks will run. If one node failed, the module will assign the mapped reduce tasks to other nodes and the transfer module can push the intermediate data to the new node without executing map tasks again. The fault tolerance module can not handle any exceptions, it's helpless if the system run into a catastrophic failure.

3.4 Data Compress Module

The data compress module lies in all the slave nodes, responsible for predicting, deciding, and compressing. This module will predict the CPU workload to decide whether compress the data in CPU or the specific hardware, and also predict the property of output data to decide which algorithm is the most appropriate compression algorithm. After predicting and deciding, the module will do the data compression using the decided algorithm. The functionalities of this module are as follows.

– Predict the CPU workload if the compressing job will be done in CPU.
– Predict the property of the output data.
– Decide whether the data compression job should be done in CPU or the specific hardware.
– Decide which is the most appropriate compression algorithm under this condition.
– Compress the data according to the decided strategy.

We find that different compression algorithm has its advantages and disadvantages after research. The data compress module implements five kinds of compression algorithms including quicklz [5], elzs, exar, snappy [6], and zlib, for different kinds of data under different conditions. First, the module predicts the CPU workload and the property of the output data. Then the module chooses the best compression algorithm. At last, do the compressing job according to the decided strategy.

The four modules remove the coupled relationship of map tasks and reduce tasks in original MapReduce, and compress the result of MapReduce jobs. It guarantees that the uncoupled MapReduce can resolve the reduce slot hoarding, underutilized network bandwidth, and inefficient storage problems.

4   Evaluations

We evaluate the model and its application effects using a micro-benchmark and a real-world example. In the micro-benchmark, we use a cluster to compare the job execution time in the uncoupled version of Hadoop with that in the original version. Our work has also been applied in a production environment of Baidu, which gives a comprehensive evaluation on the uncoupled MapReduce model.

– **Definition 1:** Wall time is the total time span from the moment a job is submitted to the moment it is completed.

- **Definition 2:** Throughput T is the number of jobs finished in a unit time interval. Suppose that N jobs are completed in a time interval t, we will get:

$$T = \frac{N}{t} \tag{5}$$

Our tests use the same workload in both the original version and the uncoupled version, so:

$$N_{original} = N_{uncoupled} \tag{6}$$

Suppose all the jobs are submitted nearly simultaneously. t_{1i} is the wall time of job i in the original version, and t_{2i} is that in the uncoupled version. Use $t_{original} = \max(\{t_{1i}\})$ and $t_{uncoupled} = \max(\{t_{2i}\})$ $(1 \leq i \leq N)$ to represent the total wall time in the original version and the uncoupled version. Then we define the throughput increment rate as I:

$$I = \frac{T_{uncoupled} - T_{original}}{T_{original}} \cdot = \frac{\frac{N_{uncoupled}}{t_{uncoupled}} - \frac{N_{original}}{t_{original}}}{\frac{N_{original}}{t_{original}}} = \frac{t_{original}}{t_{uncoupled}} - 1 \tag{7}$$

Then we define the rate of total wall time reduction as r, and the rate of job i's wall time reduction as r_i:

$$r = \frac{t_{original} - t_{uncoupled}}{t_{original}} \tag{8}$$

$$r_i = \frac{t_{1i} - t_{2i}}{t_{1i}} \tag{9}$$

The impact factor of job i, λ_i, represents the proportion of job i's wall time in all the jobs. We can get λ_i from Eq. 10:

$$\lambda_i = \frac{t_{1i}}{\sum_{i=1}^{N} t_{1i}} \tag{10}$$

The weighted wall time acceleration P represents the sum of wall time reduction rate with impact factors. It can be deduced from Eq. 11:

$$P = \sum_{i=1}^{N} \lambda_i r_i = \sum_{i=1}^{N} \frac{t_{1i}}{\sum_{i=1}^{N} t_{1i}} \frac{t_{1i} - t_{2i}}{t_{1i}} = \frac{\sum_{i=1}^{N} (t_{1i} - t_{2i})}{\sum_{i=1}^{N} t_{1i}} = 1 - \frac{\sum_{i=1}^{N} t_{2i}}{\sum_{i=1}^{N} t_{1i}} \tag{11}$$

The throughput increment rate and total wall time reduction rate are the metrics reflecting the overall performance. The weighted wall time acceleration is the metric reflecting the cumulative performance of each job.

4.1 Micro-Benchmark

The micro-benchmark is performed in a cluster with 6 nodes. The operating system is CentOS 6.1 × 86_64, and the Hadoop version is 0.19. We use gridmix [3] applications for our test. Gridmix is a set of benchmark programs for Hadoop which contains several kinds of jobs. The micro-benchmark includes 3 jobs as shown in Table 1. job1 was submitted first, then job2, and job3 at last. The interval between two adjacent jobs is 30 seconds. We divide the micro-benchmark into three parts as follows.

Table 1. The workload of the micro-benchmark

Job name	Input size	Maps	Reduces
job1	50 GB	400	200
job2	12.5 GB	100	50
job3	6.25 GB	50	25

Slot Allocation. In original MapReduce, reduce tasks of job1 occupy all the reduce slot since only job1 in the system at that time in Fig. 7. Reduce tasks of job2 and job3 can not start until one or more reduce tasks of job1 finished. As job2 and job3 are small jobs comparing to job1, when map tasks of job2 and job3 finished, no reduce tasks of job1 finished and all reduce slots were stilled occupied by job1. So it caused the reduce slot hoarding problem. As we can see in Fig. 8, the uncoupled MapReduce resolves the reduce slot hoarding problem by taking the strategy that all reduce tasks must be executed after all map tasks finished.

Network Bandwidth. Although the coupled MapReduce starts reduce tasks before all map tasks finished, most of reduce tasks have to wait due to the limitation of reduce slots. So the network load is low since only reduce tasks need large network bandwidth. As we can see in Fig. 9, network bandwidth is

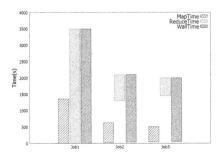

Fig. 7. Reduce slot in original MapReduce

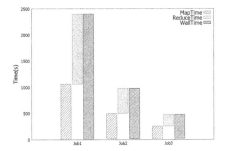

Fig. 8. Reduce slot in uncoupled MapReduce

underutilized for that the network load in reduce tasks is much higher than that in map tasks. In map tasks, the uncoupled MapReduce transfers the intermediate data to the node where the mapping reduce tasks will run on. So the uncoupled MapReduce makes full use of network bandwidth as the map tasks transfer the intermediate data and the reduce tasks write the result and replications into file system. Figure 10 shows the network traffic in uncoupled MapReduce.

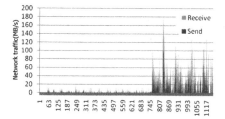
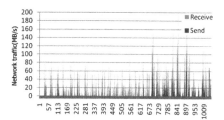

Fig. 9. Network traffic in original MapReduce

Fig. 10. Network traffic in uncoupled MapReduce

System Performance. Figure 11 shows the execution time of each job and the total completion time of all the workloads. In our test, the total time of the original version is 3500s, and that of the uncoupled version is 2620s. There is no reduce slot hoarding in the uncoupled version. The throughput increment is 34 % (I) through Eq. 7, The total wall time is reduced by 25 % (r) through Eq. 8, and the weighted wall time acceleration reaches 48 % (P) through Eq. 11. The test shows that our model can balance the network load properly and improve the system throughput. The uncoupled MapReduce makes full use of disk storage since it can reduce 70 % data volume comparing to the original MapReduce according to our benchmark.

4.2   Real-World Example

The uncoupled MapReduce implementation based on Hadoop has been deployed in a production environment of Baidu supporting some business applications. The real-world example provides strong evidence on the effects of this work.

Baidu is the biggest search engine company in China. It has tens of clusters performing Hadoop jobs for many web data processing applications, and generates more than 3 PB data volume per day [17]. Although the clusters can deal with hundreds of jobs everyday, they still meet with some problems. For example, the CPU and network bandwidth utilization rates are not high in spite of full workload. The uncoupled version of Hadoop has been deployed in a server cluster with 70 nodes, which is one of the shared Hadoop platforms for many departments. The resource scale reaches about 560 cores, 1,120 GB memory, and 770 TB storage. The operating system is Red Hat Enterprise Linux AS release 4, and the Hadoop version is 0.19. Many kinds of jobs run in this real environment,

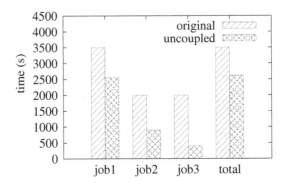

Fig. 11. Comparison of execution time in the micro-benchmark

such as log analysis, inverted index, web ranking, etc. Four kinds of jobs are used in our test, including CPU-intensive ones and I/O-intensive ones. The job information is shown in Table 2.

Table 2. The workload of the real-world example

Job name	Input size	Maps	Reduces
job1	3.4 TB	14700	1600
job2	3.2 TB	14400	1600
job3	353 GB	1700	800
job4	343 GB	1600	800

Figure 12 shows the execution time of each job and the total completion time of all the workload. In our test, the total time of the original version is 272 min, and that of the uncoupled version is 210 min. So we can get the rate of throughput increment:

$$I = \frac{t_{original}}{t_{uncoupled}} - 1 = 29.5\% \qquad (12)$$

The rate of total wall time reduction:

$$r = \frac{t_{original} - t_{uncoupled}}{t_{original}} = 22.8\% \qquad (13)$$

And the weighted wall time acceleration:

$$P = 1 - \frac{\sum_{i=1}^{4} t_{2i}}{\sum_{i=1}^{4} t_{1i}} = 26.3\% \qquad (14)$$

In addition, job3 and job4 in the original version cannot get reduce slots when their map tasks are completed. They are delayed by 51 min and 63 min respectively because of reduce slot hoarding problem.

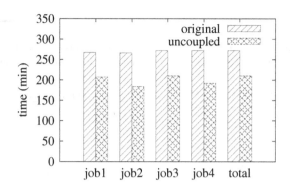

Fig. 12. Comparison of execution time in the real-world example

Figures 13, 14, 15, and 16 show the comparison of the original version and the uncoupled version in resources utilization. The workload is the above examples. In the uncoupled version, it took 3.5 h to complete all the jobs, and the last map task finished in about 2.5 h. While in the original version, it took 4.5 h to complete all the jobs, and the last map task finished in about 3.3 h.

Figure 13 illustrates the average CPU utilization rate of the cluster. The CPU utilization rate in the uncoupled version is higher than that in the original version. The reason can be explained from the map and reduce tasks respectively. In the map tasks, intermediate data is pushed and received by the corresponding servers, which results in more data preprocessing and transferring operations. In the reduce tasks, data is processed faster due to data locality.

Figure 14 shows the network load (send throughput) of the cluster. The uncoupled version uses more network bandwidth in the map tasks to transfer data. Each map will push all of its output data to its corresponding reduce tasks, instead of data being pulled partially by reduce tasks in the original version. There is little network bandwidth in reduce phase of the uncoupled version, because the reduce tasks will read data from their local disks. The peak in the end shows that HDFS uses more bandwidth to backup the output data of reduce tasks including the results and completion information of jobs.

Figures 15 and 16 show the disk I/O (write and read throughput) of the cluster. In the uncoupled version, more disk I/O workload is involved in the map task than that in the reduce task, because the map task will read data from disks and push them to the mapped nodes where the servers receive and write data. While in the original version, the average throughput of disks is relatively low. Shuffle is an ineffective and complicated phase, which has a significant impact on the execution time and system throughput.

Overall, the uncoupled version of Hadoop increases the resource utilization rates, and avoids the waste of network and disk bandwidth.

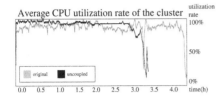

Fig. 13. Comparison of the average CPU utilization rate of the cluster

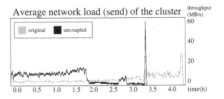

Fig. 14. Comparison of the average network load (send) of the cluster

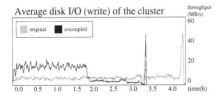

Fig. 15. Comparison of the average disk I/O (write) of the cluster

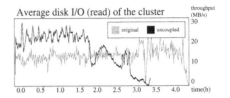

Fig. 16. Comparison of the average disk I/O (read) of the cluster

5   Related Work

There are a lot of researches on MapReduce. This section will introduce some related work. They aim at improving the system throughput and making full use of system resources.

- **The Pipeline In Hop** [11]. Hadoop Online Prototype can push the map tasks' output to reduce tasks so that there is no need to write the output as intermediate data into local disks. Then reduce tasks can go into the sorting and reducing phases without waiting for the finish of map tasks. Therefore, map and reduce tasks can work simultaneously, which can improve job performance and efficiency, and balance the network load. However, since no intermediate data stored in the local disk, the fault tolerance mechanism can not work, resulting in high cost for error recovering. In uncoupled MapReduce, we back up the intermediate data in local disk when pushing it to reduce tasks. So there is no need to re-execute map tasks when error happens in data transformation.
- **Copy-compute Splitting** [26]. This work is put forward by Facebook and includes two phases, copy and compute. The copy phase is I/O-intensive and needs to pull the map tasks' output from other nodes, while the compute phase is CPU-intensive. The reduce task consists of two kinds of tasks, the copy task and compute task, which could run simultaneously. The copy task notifies the compute task to work when it gets all the output from the map tasks. This technique can alleviate the reduce slot hoarding problem. However, it needs to set two variables, maxComputing and maxReducers. MaxComputing is the same as the reduce slot number in the original Hadoop. Limited by the number

of copy slots, it will also cause Copy Slot Hoarding problem when big jobs occupy all the copy slots.

- **Dynamic Weight Assignment** [24]. In this model, map and reduce tasks share the total number of slots. Each job can get its own slots according to its weight. At the beginning of a job, the map tasks get big weights and reduce tasks get small ones. Gradually, the map tasks get fewer slots while the reduce tasks get more slots. But the big job can still get a small number of reduce slots which can only alleviate reduce slot hoarding problem. Besides, it will also have an effect on the job priority. The job with high priority may not be scheduled in time and reduce tasks may be delayed, which can decrease the network bandwidth.

- **Weighted Shuffle Scheduling** [10]. In this model, the authors think that the shuffle phase needs to get a lot of data, which will occupy a large amount of the job operating time. Through the analysis of jobs in Facebook, it shows that shuffle can take about 33 % of the operating time in reduce tasks. Therefore, they think that shuffle plays an important role on the performance of the system. This technique can raise the efficiency of shuffle. They assign each flow (a flow is a socket connection between a map task and a reduce task) with a weight. Then the network bandwidth can be under control through this weight. However, it cannot address the reduce slot hoarding or unbalanced network load problem. The network bandwidth is still concentrated in reduce tasks.

- **Spark** [27]. MapReduce is built around an acyclic model that is not suitable for iterative algorithm and interactive data analysis. Spark is a new cluster computing framework supporting this kinds of work while retaining the scalability and fault tolerance of MapReduce. A resilient distributed dataset(RDD) is the main abstraction in Spark representing a read-only collection of objects partitioned across a set of machines that can be rebuilt if a partition is lost. This model can outperform Hadoop by 10x in iterative machine learning jobs, and can be used to interactively query a 39GB dataset with sub-second response time. However, it is not suitable for fine-grained and asynchronous update operations since Spark is a coarse-grained data parallel computing model.

- **MapReduce Energy Efficiency** [9]. Since data becomes larger and larger, the modern data center puts more and more attention on energy efficiency. Compression can reduce data dramatically and reduces the usage of disks. However, it may not improve energy efficiency since the compressing data needs additional computing. Reference [9] come up with a module to help improve the energy efficiency by 35–60% for the data read frequently on MapReduce. The module predicts the compression ratio and the visited frequency of the data to decide whether to compress. When the compression ratio is smaller than 0.2, the data will be compressed. When the compression ratio is between 0.2 and 0.4, the data will be compressed if the data will be frequently read. Under other conditions, the data will be stored and transferred directly. This model is useful in the practical application. Reference [14] also comes up with a good model to improve energy efficiency, it proves that compressing and decompressing take up 7–11% time of map tasks while 37.9–41.2% time of reduce tasks since compressing and decompressing can increase the burden of CPU. This module introduces a specific hardware to do compressing and decompressing jobs instead of CPU, which has achieved a good performance. However, the efficient of these two systems is still low since they only focus on energy efficiency when compressing without concerning the coupled relationship between maps and reduces.

– **BlobSeer** [18,19]. BlobSeer is a transparent compression module on Blob-Seer File System(BSFS) aiming at reducing the data amount and improve the throughput of the system. This module can be implemented by Hadoop and used by MapReduce. Before compressing the total data, it compresses a small part of data using sampling method to predict the compression ratio of total data and determines whether compressing is useful. Users can choose compression algorithms according their needs, but only the chosen algorithm can be used when the system is running. Experiments has been done using algorithm LZO [4] and BZIP2 [2], and result shows that BlobSeer can reduce 40% data. However, it will cause CPU overload problem that have a bad effect on other jobs. In uncoupled MapReduce, we solve this problem using a specific hardware.

6   Conclusion and Future Work

Although there are a variety of optimizations to improve the localization rate of data in the MapReduce model, data transfer is inevitable. Reduce tasks need to pull intermediate data from map tasks, which will decrease the execution efficiency of jobs. In [10], it shows that shuffle can take about 33% of the operating time in reduce tasks. Meanwhile, data transfer efficiency is very low, which is a system bottleneck [25]. Since the MapReduce model is designed to be fault-tolerant, it's costly to store the replicated file especially when the result is huge. Impacted by these facts, the original MapReduce model results in three problems: reduce slot hoarding, underutilized network bandwidth, and inefficient storage.

In this paper, we propose an uncoupled MapReduce model to resolve these problems with the aim of improving the resource utilization and system throughput. Four techniques, including weighted mapping, data pushing, partial data backup and data compression, are implemented. This work has been practiced in Baidu, the biggest search engine company in China. In a real-world application, the test shows that the throughput can be increased by 29.5%, the total wall time is reduced by 22.8%, the weighted wall time acceleration reaches 26.3%, and reduce the result data storage by 70% compared with the original version of Hadoop. The resource utilization rates of CPU, network and disk are also increased.

Two improvements are planned: (1) Design a kind of scheduler for this model which can use the cluster resources more reasonably. (2) Monitor workloads and resources dynamically, instead of setting constant slots and weights. Hopefully, our paper would assist in the study of heterogeneous resources utilization.

Acknowledgment. We would like to thank Ruijian Wang, Maosen Sun, Chen Feng, Fan Liang, and Dixin Tang from Institute of Computing Technology, Chinese Academy of Sciences for the valuable discussions. We also thank Chuan Xu, Linjiang Lian, and Meng Wang from Baidu for their assistance and support. This research is supported in part by the Hi-Tech Research and Development (863) Program of China (Grant No. 2013AA01A213, 2011AA01A203).

References

1. Apache hadoop. http://hadoop.apache.org/
2. Bzip2 compression. http://www.bzip.org/
3. Gridmix. http://hadoop.apache.org/docs/stable/gridmix.html
4. Lempel-ziv-oberhumer(lzo) compression. http://www.oberhumer.com/opensource/lzo
5. Quicklz. http://www.quicklz.com/
6. Snappy. http://code.google.com/p/snappy/
7. Cao, P., Felten, E.W., Karlin, A.R., Li, K.: A study of integrated prefetching and caching strategies. SIGMETRICS Perform. Eval. Rev. **23**(1), 188–197 (1995)
8. Chang, F., Dean, J., Ghemawat, S., Hsieh, W.C., Wallach, D.A., Burrows, M., Chandra, T., Fikes, A., Gruber, R.E.: Bigtable: A distributed storage system for structured data. ACM Trans. Comput. Syst. **26**(2), 4:1–4:26 (2008)
9. Chen, Y., Ganapathi, A., Katz, R.H.: To compress or not to compress - compute vs. IO tradeoffs for MapReduce energy efficiency. In: Proceedings of the First ACM SIGCOMM Workshop on Green Networking, Green Networking 2010, pp. 23–28. ACM, New York (2010)
10. Chowdhury, M., Zaharia, M., Ma, J., Jordan, M.I., Stoica, I.: Managing data transfers in computer clusters with orchestra. In: Proceedings of the ACM SIGCOMM 2011 Conference, SIGCOMM 2011, pp. 98–109 (2011)
11. Condie, T., Conway, N., Alvaro, P., Hellerstein, J.M., Elmeleegy, K., Sears, R.: MapReduce online. In: Proceedings of the 7th USENIX Conference on Networked Systems Design and Implementation, NSDI 2010, pp. 1–15 (2010)
12. Dean, J., Ghemawat, S.: MapReduce: simplified data processing on large clusters. In: Proceedings of the 6th USENIX Symposium on Operating Systems Design & Implementation, OSDI 2004, pp. 137–150 (2004)
13. Ghemawat, S., Gobioff, H., Leung, S.T.: The google file system. SIGOPS Oper. Syst. Rev. **37**(5), 29–43 (2003)
14. Gu, X., Hou, R., Zhang, K., Zhang, L., Wang, W.: Application-driven energy-efficient architecture explorations for big data. In: Proceedings of the 1st Workshop on Architectures and Systems for Big Data, ASBD 2011, pp. 34–40. ACM, New York (2011)
15. Gu, Y., Grossman, R.L.: Sector and sphere: Towards simplified storage and processing of large scale distributed data (2008). arXiv:0809.1181
16. Isard, M., Budiu, M., Yu, Y., Birrell, A., Fetterly, D.: Dryad: distributed data-parallel programs from sequential building blocks. In: Proceedings of the 2nd ACM SIGOPS/EuroSys European Conference on Computer Systems, EuroSys 2007, pp. 59–72 (2007)
17. Ma, R.: Introduction to part of the baidu's distributed systems. http://www.slideshare.net/cydu/sacc2010-5102684

18. Nicolae, B., Moise, D., Antoniu, G., Bouge, L., Dorier, M.: Blobseer: Bringing high throughput under heavy concurrency to hadoop map-reduce applications. In: 2010 IEEE International Symposium on Parallel Distributed Processing (IPDPS), pp. 1–11 (2010)

19. Nicolae, B.: High throughput data-compression for cloud storage. In: Hameurlain, A., Morvan, F., Tjoa, A.M. (eds.) Globe 2010. LNCS, vol. 6265, pp. 1–12. Springer, Heidelberg (2010)

20. Padmanabhan, V.N., Mogul, J.C.: Using predictive prefetching to improve world wide web latency. SIGCOMM Comput. Commun. Rev. **26**(3), 22–36 (1996)

21. Seo, S., Jang, I., Woo, K., Kim, I., Kim, J.S., Maeng, S.: Hpmr: Prefetching and pre-shuffling in shared mapreduce computation environment. In: IEEE International Conference on Cluster Computing and Workshops, CLUSTER 2009, pp. 1–8 (2009)

22. Shvachko, K., Kuang, H., Radia, S., Chansler, R.: The hadoop distributed file system. In: 2010 IEEE 26th Symposium on Mass Storage Systems and Technologies (MSST), pp. 1–10 (2010)

23. Wang, X.W., Zhang, J., Liao, H.M., Zha, L.: Dynamic split model of resource utilization in mapreduce. In: Proceedings of the Second International Workshop on Data Intensive Computing in the Clouds, DataCloud-SC 2011, pp. 21–30. ACM, New York (2011)

24. Wang, X., Zhang, J., Liao, H., Zha, L.: Dynamic split model of resource utilization in MapReduce. In: Proceedings of the 2nd International Workshop on Data Intensive Computing in the Clouds, DataCloud-SC 2011, pp. 21–30 (2011)

25. Wang, Y., Que, X., Yu, W., Goldenberg, D., Sehgal, D.: Hadoop acceleration through network levitated merge. In: Proceedings of 2011 International Conference for High Performance Computing, Networking, Storage and Analysis, SC 2011, pp. 1–10 (2011)

26. Zaharia, M., Borthakur, D., Sen Sarma, J., Elmeleegy, K., Shenker, S., Stoica, I.: Job scheduling for multi-user MapReduce clusters. Technical report UCB/EECS-2009-55, EECS Department, University of California, Berkeley (2009)

27. Zaharia, M., Chowdhury, M., Franklin, M.J., Shenker, S., Stoica, I.: Spark: cluster computing with working sets. In: Proceedings of the 2nd USENIX Workshop on Hot Topics in Cloud Computing, HotCloud 2010 (2010)

Enhanced Fast Causal Network Inference over Event Streams

Saurav Acharya$^{(\boxtimes)}$ and Byung Suk Lee

Department of Computer Science, University of Vermont, Burlington 05405, USA
{sacharya,bslee}@uvm.edu

Abstract. This paper addresses causal inference and modeling over event streams where data have high throughput, are unbounded, and may arrive out of order. The availability of large amount of data with these characteristics presents several new challenges related to causal modeling, such as the need for fast causal inference operations while ensuring consistent and valid results. There is no existing work specifically for such a streaming environment. We meet the challenges by introducing a time-centric causal inference strategy which leverages temporal precedence information to decrease the number of conditional independence tests required to establish the causalities between variables in a causal network. (Dependency and temporal precedence of cause over effect are the two properties of a causal relationship.) Moreover, we employ change-driven causal network inference to safely reduce the running time further. In this paper we present the Order-Aware Temporal Network Inference algorithm to model the temporal precedence relationships into a temporal network and then propose the Enhanced Fast Causal Network Inference algorithm for learning a causal network faster using the temporal network. Experiments using synthetic and real datasets demonstrate the efficacy of the proposed algorithms.

Keywords: Causal inference · Event streams · Temporal data

1 Introduction

In recent years, there has been a growing need for active systems that can perform causal inference in diverse applications such as health care, stock markets, user activity monitoring, smart electric grids, and network intrusion detection. These applications need to infer the cause of abnormal activities immediately from their event streams, where the event arrival may be in order (e.g., [1,2]) or out of order (e.g., [3–7]) such that informed and timely preventive measures can be taken. As a case in point, consider a smart electric grid monitoring application. The failure of a component can cause cascading failures, effectively causing a massive blackout. The identification of such cause and effect components in a timely manner enables preventive measures in the case of failure of a cause component, thereby preventing blackouts.

© Springer-Verlag Berlin Heidelberg 2015
A. Hameurlain et al. (Eds.): TLDKS XVII, LNCS 8970, pp. 45–73, 2015.
DOI: 10.1007/978-3-662-46335-2_3

Causal network, a directed acyclic graph where the parent of each node is its direct cause, has been popularly used to model causality [8–14]. There are two distinct types of algorithms for learning a causal network: score-based [8–11] and constraint-based [12–15]. Both types of algorithms are slow and, therefore, not suitable for event streams where prompt causal inference is required. Score-based algorithms perform a greedy search (usually hill climbing) to select a causal network with the highest score from a large number of possible networks. With an increase in the number of variables in the dataset, the number of possible networks grows exponentially, resulting in slow causal network inference. On the other hand, constraint-based algorithms (e.g., PC algorithm [14]) discover the causal structure via a large number of tests on conditional independence(CI). There can be no edge between two conditionally independent variables in the causal network (e.g., [16,17]). Two variables X and Y are said to be conditionally independent given a condition set S if there is at least one variable in S such that X and Y are independent (e.g., [18,19]). In a causal network of n variables, the condition set S consists of all possible 2^{n-2} combinations of the remaining $n-2$ variables, and therefore the computational complexity grows exponentially as the number of variables increases. So, the current techniques for causal inference are slow and not suitable for event streams which have a high data throughput and where the number of variables (i.e., event types) is large. Besides, these techniques perform the time-consuming causal inference computations every time a new batch of events arrives even though there may not be significant enough changes in the event stream statistic.

With these concerns, this paper describes a new time-centric causal modeling approach to speed up the causal network inference. Every causal relationship implies temporal precedence relationship (e.g., [20,21]). So, the idea is to exploit temporal precedence information as an important clue to reducing the number of required CI tests and thus maintaining feasible computational complexity. Four strategies are employed utilizing this idea to achieve fewer computations of CI tests. First, since causality requires temporal precedence, we ignore the causality test for those nodes with no temporal precedence relationship between them. Second, in the CI test of an edge, we exclude those nodes from the condition set which do not have temporal precedence relationship with the nodes of the edge; this strategy reduces the size of the condition set which is a major cause of the exponential computational complexity. Third, we perform the CI tests for weaker edges (i.e., having lower temporal strength) earlier to reduce the size of the condition set of stronger edges, thereby reducing the overall number of CI tests. The rationale for this is that weaker edges are more likely to be eliminated than stronger edges [22]. Fourth, we perform the causal inference computations only if there is a significant enough change in the temporal precedence relationships, which is a necessary condition for a change to occur in the resulting causal relationships. Such a change detection strategy helps to avoid unnecessary causal inference computations, and therefore, saves time.

Due to the reliance on the temporal precedence relationships in an event stream, events arriving *out of order* can bring ambiguities in the resulting causal directions. For instance, the precedence relationships represented in an edge and its reversed edge in a temporal network, which models the temporal precedence relationships, may not be significantly different enough to determine the edge direction. Intuitively, an undirected edge can be used to signify such an ambiguity. Thus, we propose a mechanism to decide between directed and undirected edge in the temporal network in such cases. Note that the constraint-based algorithms like the PC algorithm naturally handle out-of-order event arrivals, as these algorithms do not depend on the temporal ordering of events, and so they can provide a suitable baseline to evaluate the handling of out-of-order events in our proposed method.

The main contributions of this paper are summarized as follows. First, it presents a temporal network structure to represent temporal precedence relationships between event types and proposes an algorithm, *Order-Aware Temporal Network Inference (OATNI)*, to construct a temporal network applicable in the streaming environment. Second, it introduces a time-centric causal modeling strategy and proposes an algorithm, *Enhanced Fast Causal Network Inference (EFCNI)*, to speed up the learning of causal network. Third, it empirically demonstrates the advantages of the proposed algorithm in terms of the accuracy and speed of learning the causal network by comparing it against two state-of-art algorithms, the PC algorithm (details in Sect. 3.4) and the FCNI algorithm [23].

This paper contains the results of a comprehensive study extended from our earlier work [23]. The two algorithms in our prior work, the *Temporal Network Inference* (TNI) and the *Fast Causal Network Inference* (FCNI), are extended to the OATNI and the FECNI algorithms, respectively. Specifically, two major extensions have been made. First, the speed of the causal inference mechanism has been increased with two strategies. As the first strategy, the CI tests are performed in the increasing order of the temporal strengths of the edges in order to remove the most probable spurious edge as early as possible, which decreases the condition set size. As the second strategy, presumably unnecessary causal inference computations are avoided by determining whether the changes in temporal precedence information in the event stream are significant enough to warrant such computations. Second, the previous work made an assumption that the event stream is in order. In this paper, the support for fast causal modeling over an out-of-order event stream is added so that the temporal precedence relationships cannot be relied upon as they are. In addition to these two major extensions, the presentation has been extended throughout in many parts of the paper.

The rest of this paper is organized as follows. Section 2 reviews the existing work on causal network inference. Section 3 presents the basic concepts used in the paper. Sections 4 and 5 propose the learning algorithms of temporal network (OATNI) and faster causal network (EFCNI), respectively. Section 6 evaluates the proposed EFNCI algorithm. Section 7 concludes the paper and suggests further research.

2   Related Work

As explained earlier, there are two main approaches for causal network inference.

The first approach, score-based [8–11], performs greedy search (usually hill climbing) over all possible network structures in order to find the network that best represents the data based on the highest score. This approach, however, has two problems. First, it is slow due to the exhaustive search for the best network structure. An increase in the number of variables in the dataset increases the computational complexity exponentially. Second, two or more network structures, called the equivalence classes [24], may represent the same probability distribution, and consequently the causal directions between nodes are quite random. There is no technique for alleviating these problems in a streaming environment. Thus, score-based algorithms are not suitable for streams.

The second approach, constraint-based [12–15], does not have the problem of equivalence classes. However, it is slow as it starts with a completely connected undirected graph and thus performs a large number of CI tests to remove the edges between conditionally independent nodes. The number of CI tests increases exponentially with the increase in the number of variables in the dataset. To alleviate this problem, some constraint-based algorithms start with a minimum spanning tree to reduce the initial size of condition sets. However, this idea trades the speed with the accuracy of the causal inference. The constraint-based algorithms include IC* [12], SGS [13], PC [14], and FCI algorithm [14]. The FCI algorithm focuses on the causal network discovery from the dataset with latent variables and selection bias, which is quite different from the scope of this paper. The PC algorithm is computationally more efficient than IC* and SGS. This is why we evaluate the proposed *EFCNI* algorithm by comparing it against the PC algorithm. Like the others, the PC algorithm starts with a completely connected undirected graph. To reduce the computational complexity, it performs CI tests in several steps. Each step produces a sparser graph than the earlier step, and consequently, the condition set decreases in the next step. However, the computational complexity is still $O(n^2 \cdot 2^{n-2})$. (The details are explained in Sect. 3.4.) Therefore, the current constraint-based algorithms are not suitable for fast causal inference over streams.

There have been a number of research works on handing out-of-order event streams [3–7]. To the best of our knowledge, however, there exists no work applicable to the causal network inference. (Thus, a new approach is needed, and our approach is to allow *undirected edges* in the temporal network.) Johnson et al. [3] propose an algorithm for regular expression matching on streams with out-of-order data, which is not related to causal inference. The works by Li et al. [4] and Liu et al. [5] discuss the problem of processing event pattern queries over event streams that may contain out-of-order data. Li et al. [6] present a new architecture for stream systems for out-of-order query processing whereas Wang and Yu [7] propose algorithms for generating and matching queries to raise accuracy and shorten the response time as much as possible over out-of-order events. None of these works is related to causal network inference.

3   Basic Concepts

This section presents some key concepts needed to understand the paper.

3.1   Event Streams

An event stream in our work is a sequence of continuous and unbounded timestamped events. An event refers to any action that has an effect and is created by an event owner. One event can trigger another event in chain reactions. Each event instance belongs to one and only one event type which is a prototype for creating the instances. Two event instances are related to each other if they share common attributes such as event owner, location, and time. We call these attributes *common relational attributes (CRAs)*.

In this paper we denote an event type as E_j and an event instance as e_{ij}, where i indicates the CRA and j indicates the event type.

Example 1. Consider a diabetic patient monitoring system in a hospital. Each patient is uniquely identifiable, and each clinical test or measurement of each patient makes one event instance. For example, a patient is admitted to the hospital, has their blood pressure and glucose level measured, and takes medication over a period of time. This creates the instances of the above event types as a result. Typical event types from these actions include hypoglycemic-symptoms-exists, blood-glucose-measurement-decreased, increased, regular-insulin-dose-given, etc. Note that the patient ID is the CRA, as the events of the same patient are causally related.

To facilitate the handling of events in a streaming environment, we use a time-based window over the stream. Typically, the application offers a natural observation period (e.g., hour) that makes a window. The causal relationship is only possible between events with the same CRA. Therefore, the events in a window are arranged by CRA and then ordered by the timestamp as they arrive, producing a *partitioned window* as a result. Figure 1 illustrates it. (We refer to the *partitioned window* simply as the *window* for the rest of the paper.)

With the arrival of a new batch of event instances, we augment each partition in the new window by prefixing it with the last instance of the partition with the

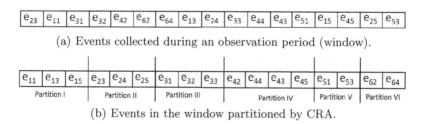

(a) Events collected during an observation period (window).

(b) Events in the window partitioned by CRA.

Fig. 1. Partitioned window of events

same CRA value in the previous window. This is necessary in order to identify the related instances that are separated into the two consecutive batches.

We support event streams which may be in order or out of order. An event stream is said to be in order if and only if every event in every partition arrives in the same temporal order as it was created. In other words, the stream is out of order if any event in any partition arrives in a different temporal order than it was created. The degree of out-of-order, d_{oo}, is given as

$$d_{oo} = \frac{\sum_{k=1}^{n_p} o_k}{n_{ins}} \tag{1}$$

where n_p is the number of partitions, o_k is the number of out-of-order events in the k-th partition and n_{ins} is the total number of events in all partitions. Note that d_{oo} is zero for in-order event streams.

3.2 Causal Networks

Causal network is a popularly used data structure for representing causality [8–11,25]. It is a graph $G = (N, \xi)$ where N is the set of nodes (representing event types) and ξ is the set of edges between nodes. For each directed edge, the parent node denotes the cause, and the child node denotes the effect.

Consider the event stream of Fig. 1. The causal relationships among the event types in the stream may be modeled as a causal network like the one shown in Fig. 2.

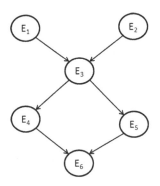

Fig. 2. Causal network

The joint probability distribution of a set of n event types $\mathbf{E} \equiv \{E_1, ..., E_n\}$ in a causal network is specified as

$$P(\mathbf{E}) = \prod_{i=1}^{n} P(E_i | \mathbf{Pa_i})$$

where $\mathbf{Pa_i}$ is the set of the parent nodes of event type E_i.

3.3   Conditional Mutual Information

A popular approach for testing the conditional independence, with respect to the joint probability P, of two random variables X and Y given a subset S of random variables is conditional mutual information(CMI) (e.g., [15,26]). CMI gives the strength of dependency between variables in a measurable quantity, which helps to identify strong and weak causal relationships in the final causal network.

To test whether X and Y are conditionally independent given S, we compute the conditional mutual information $I_{MI}(X, Y|S)$ as

$$I_{MI}(X, Y|S) = \sum_{x \in X} \sum_{y \in Y} \sum_{s \in S} p_{X,Y,S}(x, y, s) log_2 \frac{p_{X,Y|S}(x, y|s)}{p_{X|S}(x|s) p_{Y|S}(y|s)}$$

where p is the probability mass function calculated from the frequencies of variables.

We only keep the record of these frequencies, not the whole events, by updating them as a new batch of events arrives. Consequently, the independence test procedure is incremental in our case.

It is said that two variables X and Y are independent when $I_{MI}(X, Y|S) = 0$; otherwise, they are dependent. However, this presents us with the risk of spurious relationships due to weak dependencies (we cannot assume $I_{MI}(X, Y|S) = 10^{-5}$ and $I_{MI}(X, Y|S) = 10$ provide the same degree of confidence in the dependency). With an increase in the value of $I_{MI}(X, Y|S)$, the dependency between the variables X and Y grows stronger. Therefore, to prune out weak dependencies, we need to set a threshold value of mutual information below which we ignore the evidence as weak. To do so, we relate CMI with G^2 test statistics [14,27] as below, where N_s is the sample size.

$$G^2(X, Y|S) = 2 \cdot N_s \cdot log_e 2 \cdot I_{MI}(X, Y|S)$$

G^2 follows the χ^2 distribution [28], with the degree of freedom df equal to $(r_x - 1)(r_y - 1) \prod_{s \in S} r_s$, where r_x, r_y, and r_s are the number of possible distinct values of X, Y, and S, respectively. So, we use χ^2 test, which provides a threshold based on df and significance level α, to validate the dependency result. We set α to the universally accepted value of 95 %.

3.4   The PC Algorithm

The key idea of the PC algorithm [14] is that a causal network has an edge between two nodes, X and Y, if and only if X and Y are not independent given any of the condition subsets of the remaining neighbor nodes [16,17]. Algorithm 1 outlines the PC algorithm.

The algorithm has two parts. In the first part, the algorithm learns the topology of the causal network (Lines 1– 18). It starts with a completely connected undirected graph G of n nodes. Two sets are used for bookkeeping – Neighbors(G,X) and SepSet(X,Y). Neighbors(G,X) gives the set of nodes adjacent

---

Algorithm 1. PC algorithm

Require: *Window W*
 1: Construct the completely connected undirected graph G of all nodes;
 2: **for** each node X in G **do**
 3:    Initialize *Neighbors(G,X)* as the set of nodes adjacent to X in G;
 4: **end for**
 5: **for** each pair of nodes X and Y in G **do**
 6:    Initialize an empty set *SepSet(X,Y)* as the set of nodes that causes independence between X and Y in G;
 7: **end for**
 8: $k \leftarrow 0$;
 9: **repeat**
10:    **repeat**
11:       Select any edge $X - Y$ such that $|Neighbors(G, X) \backslash Y| \geq k$;
12:       **repeat**
13:          Select any subset S of $Neighbors(G, X) \backslash Y$ such that $|Neighbors(G, X) \backslash Y| = k$;
14:          If X and Y are independent given S, remove $X - Y$ from G, remove Y from Neighbors(G,X), remove X from Neighbors(G,Y), and add S to SepSet(X,Y) and SepSet(Y,X);
15:       **until** every subset S of $Neighbors(G, X) \backslash Y$ such that $|Neighbors(G, X) \backslash Y| = k$ has been selected.
16:    **until** every edge $X - Y$ such that $|Neighbors(G, X) \backslash Y| \geq k$ has been selected.
17:    k = k + 1;
18: **until** no edge $X' - Y'$ satisfies $|Neighbors(G, X') \backslash Y'| \geq k$.
19: **for** each triplet of nodes X, Y, Z such that the edge $X - Y$ and $Y - Z$ exists in G but not $X - Z$ **do**
20:    Orient $X - Y - Z$ as X→Y←Z if and only if SepSet(X,Z) does not contain Y;
21: **end for**
22: **repeat**
23:    **If** there exists $X \rightarrow Y$ and $Y - Z$, but not $X - Z$, **then** orient $Y - Z$ as $Y \rightarrow Z$;

24:    **If** there exists $X - Y$ and a directed path from X to Y, **then** orient $X - Y$ as $X \rightarrow Y$;
25: **until** no edge can be oriented.

---

to X, and *SepSet(X,Y)* gives the set of nodes which causes X and Y to be conditionally independent. Initially, *Neighbors(G,X)* has the remaining $n - 1$ nodes and *SepSet(X,Y)* is empty. Then, the CI tests are performed between every pair of nodes that have an edge between them to determine whether they are conditionally independent of any other nodes in G. The edge between the conditionally independent nodes is removed. To ensure that all possible combination of nodes are considered in the condition set, the algorithm starts with an empty condition set and ends with the condition set with the maximum possible number of nodes. That is, in the algorithm k refers to the number of nodes in the condition set and is initially set to 0 to denote an empty condition set, and then k is gradually increased (by 1 at each iteration) and the CI tests are performed between every pair of nodes with the condition set of size k. This process is repeated until there is no edge left in G whose condition set size is greater than k. Eventually, an undirected network is obtained where an edge between two nodes denotes that these nodes are not conditionally independent in the presence of any of the other $n - 2$ nodes.

In the second part, the undirected network topology is assigned causal directions (Lines 19 – 25). It is done in three steps. First, if there are two edges $X - Y$ and $Y - Z$ but not $X - Z$ and $SepSet(X,Z)$ does not contain Y, then $X - Y - Z$ is assigned the edge orientations $X \rightarrow Y \leftarrow Z$. The reason is that X and Z are dependent given Y, as the absence of Y in $SepSet(X,Z)$ indicates that X and Z are not conditionally independent given Y [29,30]. Second, if there are edges $X \rightarrow Y$ and $Y - Z$ but not $X - Z$, then $Y - Z$ is oriented as $Y \rightarrow Z$. The absence of an edge between X and Z means that X and Z are not dependent. A directed edge from X to Y and Y to Z with no edge between X and Z makes X and Z independent of Y [29,30]. Thus, the edge between Y and Z is oriented as $Y \rightarrow Z$. Third, if there are edges $X - Y$ and a directed path from X to Y through any number of nodes, then $X - Y$ is oriented as $X \rightarrow Y$. This is necessary to make the graph acyclic, as the edge direction $X \leftarrow Y$ would make the graph cyclic.

4   Learning Temporal Precedence Relationships

In this section, we describe an incremental approach to model temporal precedence relationships from time-stamped events into a *temporal network*.

4.1   Temporal Network Model

A temporal network is a network of nodes representing event types where an edge between two nodes represents the temporal precedence relationship between them. To determine when an edge should be added in a temporal network, a measure providing an evidence of temporal precedence between the event types should be defined. The evidence we use is the frequency of the observation of an instance of E_j following an instance of E_i. We call this the *precedence frequency*.

Definition 1 (Precedence frequency). *The precedence frequency f_{ij} between two event types E_i and E_j is the total number of observations in which an event of type E_i precedes an event of type E_j over all partitions in the partitioned window.*

$$f_{ij} = \sum_{k=1}^{n_p} n_{(e_{ki}, e_{kj})}$$

where n_p is the number of partitions and $n_{(e_{ki}, e_{kj})}$ is the number of observations in which an event of type E_i precedes an event of type E_j in the k-th partition.

$\square$

In our prior work [23], we assume that the event stream is in order and the temporal network from such an event stream is directed and acyclic. However, in an event stream with out-of-order event arrivals, the reliance on the temporal order for a definite temporal edge direction between event types may lead to ambiguous scenarios. For instance, the precedence frequencies between two event types E_i and E_j of an edge $E_i \rightarrow E_j$ and its reversed edge $E_i \leftarrow E_j$ may not differ significantly enough to determine an edge direction between them. An undirected edge $E_i - E_j$ is warranted to reflect such ambiguity. Thus, we support undirected edges as well as directed edges in the temporal network model. A directed edge between two nodes reflects a strong temporal precedence relationship between them, whereas an undirected edge reflects an ambiguous temporal precedence relationship between them. A threshold called the *temporal confidence* (θ) is used to select between directed and undirected edges, as presented in Rule 1 below.

Rule 1. Temporal edge direction selection
Suppose the precedence frequencies of an edge $E_i \rightarrow E_j$ and its reversed edge $E_i \leftarrow E_i$ are f_{ij} and f_{ji} such that either $f_{ij} > 0$ or $f_{ji} > 0$, respectively. Then, the edge direction between these two event types E_i and E_j (i.e., $\Xi(E_i, E_j)$) is selected as follows.

$$\Xi(E_i, E_j) = \begin{cases} E_i \rightarrow E_j & \text{if } \frac{f_{ij}-f_{ji}}{f_{ij}+f_{ji}} > \theta \\ E_i \leftarrow E_j & \text{if } \frac{f_{ji}-f_{ij}}{f_{ij}+f_{ji}} > \theta \\ E_i - E_j & \text{if } |\frac{f_{ij}-f_{ji}}{f_{ij}+f_{ji}}| \leq \theta \end{cases}$$

$\square$

Given a temporal network, we define the edge strength, called the *temporal strength*, as follows.

Definition 2 (Temporal strength). *Consider an edge $E_i \rightarrow E_j$ ($i \neq j$) in a temporal network of n event types. Let f_{ij} be the precedence frequency from the event type E_i to the event type E_j. Then, we define the* temporal strength, s_{ij}, *of the edge $E_i \rightarrow E_j$ as*

$$s_{ij} \triangleq \frac{f_{ij}}{\sum_{k=1}^{n} f_{ik}}$$

$\square$

That is, the temporal strength of $E_i \rightarrow E_j$ is the precedence frequency of (E_i, Ej) relative to the total precedence frequency over all children nodes of E_i.

4.2   Order-Aware Temporal Network Inference Algorithm

The idea behind the *OATNI* algorithm is to collect events from an event stream in a *window* and then use temporal precedence information from the sequence of event pairs in the window to construct a temporal network at the event type level. The overall algorithm is centered on a *frequency matrix*, which is initially empty (i.e., all zero elements) and updated with each new batch of events.

The algorithm has two steps for each window, as outlined in Algorithm 2.

Algorithm 2. Order-Aware Temporal Network Inference (OATNI)

Require: *Edgeless network structure TN, Event stream(s) S, Temporal confidence θ*
1: Initialize an empty frequency matrix *FM*, an empty strength matrix *SM*, two empty buffers B_p and B_c (used to store "parent" events and "child" events, respectively);
2: **for** each window W in S **do**
3:   **for** each partition P (corresponding to CRA a) in W **do**
4:     **for** i = 1 to $t_n - 1$ where t_n is the number of unique timestamps in P **do**
5:       Clear B_p and B_c;
6:       Insert all events with timestamp t_i and t_{i+1} into B_p and B_c, respectively;
7:       **for** each event instance e_{ap} in B_p **do**
8:         **for** each event instances e_{ac} in B_c **do**
9:           **if** type(e_{ac}) $\neq$ type(e_{ap}) {*//There cannot be causal relationships between events of the same type.*} **then**
10:             Increase the frequency of element $f_{type(e_{ap}),type(e_{ac})}$ in *FM* by 1;
11:           **end if**
12:         **end for**
13:       **end for**
14:     **end for**
15:   **end for**
16: **for** each pair of elements f_{ij} and f_{ji} such that $f_{ij} > 0$ or $f_{ji} > 0$ in *FM* **do**
17:   $s_{ij} \leftarrow 0$, $s_{ji} \leftarrow 0$;
18:   **if** $\frac{f_{ij}-f_{ji}}{f_{ij}+f_{ji}} > \theta$ **then**
19:     Add an edge $E_i \rightarrow E_j$ in *TN* and set its strength to $s_{ij} = \frac{f_{ij}}{\sum_{k=1}^{n} f_{ik}}$;
20:   **else if** $\frac{f_{ji}-f_{ij}}{f_{ij}+f_{ji}} > \theta$ **then**
21:     Add an edge $E_i \leftarrow E_j$ in *TN* and set its strength to $s_{ji} = \frac{f_{ji}}{\sum_{k=1}^{n} f_{jk}}$;
22:   **else if** $\frac{|f_{ij}-f_{ji}|}{f_{ij}+f_{ji}} \leq \theta$ **then**
23:     Add an edge $E_i - E_j$ in *TN* and set the strengths to $s_{ij} = \frac{f_{ij}}{\sum_{k=1}^{n} f_{jk}}$ and $s_{ji} = \frac{f_{ji}}{\sum_{k=1}^{n} f_{jk}}$;
24:   **end if**
25: **end for**
26: **end for**

---

1. Update the frequency matrix *FM* by observing the precedence relationships of event pairs in the partitioned window (Lines 3–15). An element f_{ij} in *FM* reflects the total number of times events of type E_i precede events of type E_j ($i \neq j$). Each time an event pair (e_{oi}, e_{oj}) is observed in the event stream such that e_{oi} precedes e_{oj}, increase the value of f_{ij} by 1.
2. Determine the edges of the temporal network using *FM* (Lines 16–25). For each pair of nodes, add an edge according to Rule 1. For a directed edge added, e.g., $E_i \rightarrow E_j$, calculate its temporal strength and store it in s_{ij} of SM. For an undirected edge added, e.g., $E_i - E_j$, calculate the temporal strengths of both directions and store them in s_{ij} and s_{ji} of SM.

Note that when the events arrive in order, the *OATNI* algorithm reduces to the *TNI* algorithm by setting the threshold θ to 0 so that Rule 1 always chooses between the first two cases.

5   Learning Causal Network in Reduced Time

In this section, we describe a new approach which reduces the number of CI tests needed to infer the causal structure, thereby speeding up the learning process. Specifically, we explain the key ideas employed and discuss the concrete algorithm. We then prove the correctness of the algorithm and analyze its computational complexity.

5.1   Key Ideas

Given that the key approach is to exploit temporal precedence relationships to learn the causal network, there are a number of ideas employed to reduce the causal network construction time. We begin by proposing some preliminary lemmas.

Lemma 1. *A CI test between two event types with no temporal precedence relationship is unnecessary.*

Proof. Two event types can a have causal relationship only if they have a temporal precedence relationship. Therefore, it is not necessary to perform a CI test (for detecting causality) between two event types which are not temporally related. □

Lemma 2. *Event types which do not have temporal precedence relationships with either of the two event types being tested for causality are not needed in the condition set of the CI test.*

Proof. Consider two event types, E_i and E_j, tested for causality, and consider another event type E_k ($k \neq i, j$). E_k can causally influence E_i (or E_j) only if E_k has a temporal precedence relationship with E_i (or E_k). Therefore, the CI tests between E_i and E_j can safely exclude from the condition set those event types (i.e., E_k) which are not temporally related to either of them. □

Based on Lemma 1, the CI tests are performed only for the edges in the temporal network. That is, not every possible edges are considered for the CI tests leading to a reduced number of CI tests. Moreover, since the size of the condition set contributes to the number of CI tests exponentially, we use Lemma 2 to reduce the condition set size by including only those event types which have temporal relationships, hence possibly causal relationships, to the event types being tested.

We employ another idea to speed up the network inference further, based on Lemma 3 below.

Lemma 3. *The number of CI tests performed in the causal network inference decreases if the CI tests between event types are performed in an increasing order of their temporal strength.*

Proof. Event types with weaker temporal strengths between them have higher likelihood of being conditionally independent than those with stronger temporal strengths. Therefore, if the CI tests are performed between event types with the lowest temporal strength first, then the initial causal network becomes sparser faster and, consequently, the condition sets for the CI tests between event types become smaller faster. This leads to the reduction in the total number of CI tests performed through the causal network inference.   □

Evidently, the reduction in the number of CI tests brings the reduction of running time.

Further, we employ the idea of reducing the overhead of causal network inference by performing it only when there are significant enough changes in temporal precedence relationships in the event stream. The rationale for this is that the causal network tends to absorb changes in the temporal network until the changes are significant enough. We introduce the temporal *precedence probability* as the measure to normalize the precedence frequencies between event types. The changes in the precedence probabilities give a normalized measure of the changes that have occurred in the temporal network since the last batch of events in the stream.

Definition 3 (Precedence probability). *The precedence probability p_{ij} between two event types E_i and E_j is defined as the ratio of f_{ij} and the summation of all precedence frequencies.*

$$p_{ij} = \frac{f_{ij}}{\sum_{x=1}^{n} \sum_{y=1}^{n} f_{xy}}$$

□

Let PM@t_i and PM@t_{i+1} be the matrices representing the precedence probabilities at the timestamps t_i and t_{i+1}, respectively. Then, the measure of change in the precedence information, called *precedence change* (C_p), is calculated as follows.

$$C_p = \sum_{x=1}^{n} \sum_{y=1}^{n} |p_{xy}@t_{i+1} - p_{xy}@t_i|$$

where p_{xy} is the element at the position (x, y) (i.e., event types (E_x, E_y)) in PM. Given this change measure, we update the causal network only if the calculated C_p exceeds a certain threshold, called the *precedence change confidence* (δ).

Algorithm 3. Enhanced Fast Causal Network Inference (EFCNI)

Require: *Window W, Precedence change confidence δ, Edgeless causal network G.*

1: Run the *OATNI* algorithm and initialize $G = (N, \xi)$ with the learned temporal network; {N and ξ are the set of nodes and the set of edges, respectively.}

2: Calculate C_p. If $C_p < \delta$, then exit; {Stop if there is no significant change in the event stream.}

3: Sort the edges in ξ in the increasing order of their temporal strength.

4: **for** each edge $(E_i, E_j) \in \xi$ **do**

5:    *independent* = $IsIndependent(E_i, E_j, \phi)$, where ϕ is the empty set; {$IsIndependent(E_i, E_j, S)$ calculates $I_{MI}(E_i, E_j | S)$ for CI test.}

6:    **if** *independent* is true **then**

7:       Remove (E_i, E_j) from ξ;

8:    **end if**

9: **end for**

10: $k \leftarrow 0$;

11: **repeat**

12:    **for** each edge $(E_i, E_j) \in \xi$ **do**

13:       Construct a set of condition sets, Z, each of cardinality k from the *parents* of E_j excluding E_i;

14:       **repeat**

15:          Select any subset S from Z;

16:          *independent* = $IsIndependent(E_i, E_j, S)$;

17:          Remove S from Z;

18:       **until** Z is empty or *independent* is true

19:       **if** *independent* is true **then**

20:          Remove (E_i, E_j) from ξ;

21:       **end if**

22:    **end for**

23:    $k = k + 1$;

24: **until** there is no E_j in any edge $(E_i, E_j) \in \xi$ with k incident edges.

5.2   Enhanced Fast Causal Network Inference Algorithm

The algorithm has four steps, as outlined in Algorithm 3.

1. The first step (Line 1) learns a temporal network by running the *OATNI* algorithm. The temporal network, which can have both directed and undirected edges, is set as the initial causal network.
2. The second step (Line 2) checks if there has been a significant enough change in the temporal precedence statistic (i.e., C_p) in the event stream from the last observation period, and stops if not.
3. The third step (Line 3) sorts the edges of the initial causal network in the increasing order of their temporal strength.
4. The fourth step (Lines 4–24) constructs the final causal network by pruning out the edges between independent nodes. CI tests are performed on every edge between adjacent nodes in the initial causal network to verify dependency between them. Conditionally independent nodes are considered spurious, and hence the edge between them is removed.

The main difference from the PC algorithm is the manner in which the CI tests are performed. In the PC algorithm, the condition set S for an edge $E_i - E_j$ (*undirected*) considers the neighbors of both E_i and E_j whereas in the EFCNI algorithm, the condition set S for an edge $E_i \rightarrow E_j$ (*directed*) needs to consider only the parents of E_j. (E_j is independent of the parents of E_i that do not have edge to E_j.) Consequently, fewer CI tests are needed. In addition, note that the *EFCNI* algorithm reduces to the *FCNI* algorithm by omitting the second and the third steps.

5.3   Correctness of the Algorithm

To prove the correctness of the algorithm, it suffices to prove the correctness of our approach which starts with a temporal network as the initial causal network and removes edges through CI tests on them. We show the correctness as follows. First, a temporal precedence relationship is a necessary condition for inferring causality [20]. Therefore, causal relationship subsumes temporal precedence relationship, that is, the causal network is a subgraph of the temporal network (Lemma 1). Second, a causal network should satisfy the *Causal Markov Condition (CMC)* [13,16,31] where for every node X in the set of nodes N, X is independent of its non-descendants excluding its parents (i.e., $N \backslash (Descendants(X) \cup Parents(X))$) given its parents. In a temporal network of vertex (or node) set N, a node is temporally independent, and therefore causally independent, of all its non-descendants (except its parents) given its parents (Lemma 2).

5.4   Complexity Analysis

Given n nodes, the computational complexity of the EFCNI algorithm is $O(n^2 \cdot 2^{n-2})$ in the worst case and $O(n)$ in the best case.

Proof. The computational complexity of the EFCNI algorithm is governed by the total number of possible CI tests which is calculated by summing up the number of CI tests involving each edge. In the worst case, the number of edges in the network is that of a completely connected graph and all edges are undirected. The number of edges in a completely connected graph of n nodes is $\frac{n(n-1)}{2}$. For every edge between two nodes, the remaining $n - 2$ nodes are considered in the condition set, as the graph is completely connected and undirected. Therefore, to test conditional independence between a pair of nodes in an edge, there are 2^{n-2} CI tests to perform. Consequently, the total number of CI tests for all edges is $\frac{n(n-1)}{2} \cdot 2^{n-2}$, resulting in the computational complexity of $O(n^2 \cdot 2^{n-2})$.

In the best case, the initial causal network (i.e., temporal network) is a directed linear graph and the number of edges is the minimum (i.e., $n - 1$). In such a graph, there are $n - 2$ edges with one incoming edge to either of the nodes and one edge with no incoming edge to either of the nodes. For the edges with one incoming edge, the condition set size is one, and therefore there are two CI tests to perform. For $n - 2$ such edges, there are $2n - 4$ CI tests. For the remaining one edge with no incoming edge to either of the nodes, there is only one CI test to perform. Therefore, there are $2n - 3$ CI tests to perform in the best case, resulting in the computational complexity of $O(n)$.   □

The computational complexity of the PC algorithm is $O(n)$ in the best case and $O(n^2 \cdot 2^{n-2})$ in the worst case [14]. Note that, while the computational complexities are the same, the EFCNI algorithm starts with a sparse network as the use of temporal precedence relationships removes many of the edges. So, it starts closer to the best case. In contrast, the PC algorithm always starts with a completely connected dense network. So, it starts from the worst case. As a result, in practice the EFCNI algorithm shows significant improvement in runtime over the PC algorithm.

The computational complexity of the FCNI algorithm is $O(n)$ in the best case and $O(n \cdot 2^{n-2})$ in the worst case [23]. FCNI's worst case computation complexity is lower than that of EFCNI by a factor of n. However, unlike the EFCNI algorithm, the FCNI algorithm is not suitable for out-of-order event streams. Moreover, as mentioned earlier, the EFCNI algorithm reduces to the FCNI algorithm when the events are in order and θ is zero. As a result, in practice the EFCNI algorithm is at least as fast and accurate as the FCNI algorithm when the events are in order and preserves the accuracy in the face of out-of-ordered events, compromising the runtime to some extent as an increasing number of events arrive out of order.

6   Performance Evaluation

We conducted experiments to compare the proposed EFCNI algorithm against the FCNI and the PC algorithms. There are three sets of experiments – first in terms of the accuracies of the resulting causal networks, second the running time, and third the number of CI tests required. In each set of experiments, we consider both the cases of stream being in order and out of order and also see the effect of the EFCNI's change-driven causal network construction strategy by comparing it with FCNI and PC when there are changes in the event stream statistic. Section 6.1 describes the experiment setup, Sect. 6.2 explains the datasets used, and Sect. 6.3 presents the experiment results.

6.1   Experiment Setup

6.1.1   Evaluation Metrics

The evaluation metrics are the speed of the causal network generation and the accuracy of the generated causal network. The running time is the CPU time, and the number of performed CI tests affects the speed. The accuracy is evaluated by examining how closely the constructed causal network structure resembles the target causal network. For this, we adopt the *structural Hamming distance* proposed by Tsamardinos et al. [32] as the measure. The nodes (i.e., event types) are fixed as given to the algorithms, and therefore the network structures are compared with respect to the edges between nodes. There are three kinds of possible errors in the causal network construction: reversed edges, missing edges, and spurious edges. We use the number of erroneous edges of each kind as the evaluation metric.

6.1.2   Platform

The experiments are conducted on RedHat Enterprise Linux 5 operating system using Java(TM) 2 Runtime Environment–SE 1.5.0_07 in Vermont Advanced Computing Core (VACC) cluster computers.

6.2   Datasets

Experiments are conducted using both synthetic and real datasets.

Synthetic Datasets. A synthetic dataset is reverse-engineered from a target causal network. Given control parameters in Table 1, the idea is to generate a random causal network, and then convert the causal network to an event stream which reflects the underlying probability distribution of the causal network. Specifically, there are three steps. First, N_{ET} nodes are created and edges are added randomly, and random conditional probabilities are assigned to each edge. Each node can have up to Max_{NC} edges from cause nodes and up to Max_{NE} edges to effect nodes. (We set both Max_{NC} and Max_{NE} to 3 for the experiments presented here.) Second, a joint probability distribution (JPD) table is built from the conditional probabilities assigned to edges of the target causal network. The rows of the JPD table collectively cover all event sequences possible, while each row has its own probability. Third, the probability for each row in the JPD table is multiplied by N_O to calculate the number of repetitions of that event sequence in the dataset. We assume that the event owner is the CRA for the dataset.

Table 1. Control parameters for synthetic event stream generation

Parameter	Meaning
N_O	Number of event owners (with unique ID)
N_{ET}	Number of event types (i.e., nodes)
Max_{NC}	Maximum number of cause events (parents)
Max_{NE}	Maximum number of effect events (children)

The size of a JPD table grows exponentially with N_{ET} and therefore we use parallel processing for the event stream generation. The JPD table is divided into multiple partitions and the dataset is created by running parallel processes over each of these partitions. The dataset is thus represented by a collection of files in which the events are shuffled according to the owner ID while preserving the temporal order.

There are five cases of datasets, DS1 through DS5, according to the number of nodes in the represented target causal networks (see their profiles in

Table 2). The target causal networks have 4, 8, 12, 16 and 20 nodes, respectively. They are created with 1, 2, 16, 64 and 512 parallel processes, respectively, thus consisting of 1, 2, 16, 64 and 512 files, respectively. Each row of a synthetic dataset represents one event instance. To obtain out-of-order event streams, each case of datasets is shuffled randomly up to the required degree of out-of-order (see Eq. 1). Changes in the event stream statistic is achieved by altering the precedence frequencies of events. Specifically, we generate six batches of the event stream for six observation points (t_1 through t_6) with the C_p values of 14 %, 16 %, 4 %, 6 %, 10 %, and 12 %, respectively, for each case of the datasets. The six batches are equal-sized for each dataset specified in Table 2, so the number of instances in a single batch is 2521, 20745, 528874, 8374548, and 85161947 for the dataset D_1, D_2, D_3, D_4, D_5, and D_6, respectively.

Table 2. Profiles of the five synthetic datasets

Dataset	N_{ET}	N_{edges}	N_O	N_ins
DS1	4	4	5000	15128
DS2	8	15	30000	124475
DS3	12	22	500000	3173246
DS4	16	39	6553600	50247293
DS5	20	49	52428800	510971687

(N_{edges} is the number of *actual* edges in the network. N_{ins} is the average number of event instances in the datasets of each case.)

Real dataset. The real dataset D_R contains diabetes lab test results [33] of 70 different patients over a period ranging from a few weeks to a few months. The dataset has a total 28143 records, about 402 records for each patient. Each record has four fields – date, time, test code, test value. The clinical data of a patient is independent of other patients. Therefore, the patient ID is the *CRA* for this dataset. There are 20 different test codes appearing in the file (shown in the left column of Table 3) from which we define event types of interest (shown in the right column of Table 3).

6.3   Experiment Results

We run the EFCNI, FCNI and PC algorithms over each type of the five synthetic datasets and the real dataset. We present our evaluation in each of the three sets of experiments. First, we evaluate the accuracy of the generated causal networks against the target causal network and determine how closely they resemble the true causal network. Specifically, we count the number of spurious edges, the number of missing edges, and the number of reversed edges. Second, we evaluate the running time (CPU time), and third, we evaluate the number of CI tests performed. We show that reducing the number of CI tests is the key to reducing

Table 3. Event types defined from the diabetes dataset

Test Code	Event Type
Regular insulin dose	Regular-insulin-dose-given(RIDG)
NPH insulin dose	NPH-insulin-dose-given(NIDG)
UltraLente insulin dose	UltraLente-insulin-dose-given(UIDG)
Unspecified BGM*	
Pre-breakfast BGM*	
Post-breakfast BGM*	Blood-glucose-measurement-
Pre-lunch blood BGM*	increased(BGMI)
Post-lunch BGM*	Blood-glucose-measurement-
Pre-supper BGM*	decreased(BGMD)
Post-supper BGM*	
Pre-snack BGM*	
Hypoglycemic symptoms	Hypoglycemic-symptoms-exist(HSE)
Typical meal ingestion	Typical-meal-ingested(TMI)
More than usual meal ingestion	More-than-usual-meal-ingested(MTUMI)
Less than usual meal ingestion	Less-than-usual-meal-ingested(LTUMI)
Typical exercise activity	Typical-exercise-taken(TET)
More than usual exercise activity	More-than-usual-exercise-taken(MTUET)
Less than usual exercise activity	Less-than-usual-exercise-taken(LTUET)

(Note BGM*: blood glucose measurement)

the running time of causal network inference. In each set of experiments, the evaluation covers the scenarios of the event stream being in order and out of order, and, additionally, the scenario of the event stream statistic changing. For the latter scenario, the value of the precedence change confidence δ is set to 9 % for all synthetic datasets (DS_1 through DS_5). For the experiments involving in-order event streams, the temporal precedence confidence θ is set to zero (so EFCNI reduces to FCNI) and, for the experiments involving out-of-order event streams, it is set to 24.80 %, 17.23 %, 18.19 %, 21.97 %, 26.50 % (each determined after training from 70 % of the data) for all synthetic datasets. The experiment is repeated ten times for each dataset (DS_1 through DS_5 and D_R) to calculate the average.

6.3.1 Comparison of the Accuracies of the PC, FCNI, and EFCNI Algorithms

6.3.1.1 *When the events arrive in order*

Table 4 presents the number of erroneous edges in the causal network produced by the PC, FCNI, and EFCNI algorithms. The results show that the accuracy of the causal network from the EFCNI algorithm is similar to that of the FCNI and PC algorithms. First, the number of missing and the number of spurious edges are comparable among all three algorithms. This is due to the reliance of the three algorithms on the same test statistics (CMI in our case) to infer the independence of two event types. Additionally, each number is the same between EFCNI and FCNI because EFCNI reduces to FCNI. Second, the number of reversed edges is zero for both the FCNI and EFCNI algorithms. Clearly the FCNI and EFCNI algorithms, through the temporal network, are much better at determining the correct causal edge directions. It is because of the fact that the cause always precedes its effect is embodied in the temporal precedence relationship. Overall, the results show that, when the event stream is in order, the EFCNI algorithm produces the same topology as the FCNI algorithm and almost the same topology as the PC algorithm, while the accuracy of the causal directions in the EFCNI algorithm remains the same as the FCNI algorithm and is improved over the PC algorithm.

Table 4. Number of erroneous edges in an in-order event stream

Type of Erroneous Edges	Algorithm	Dataset					
		DS_1	DS_2	DS_3	DS_4	DS_5	D_R
Missing	PC	0	0	0	0	1	1
	FCNI	0	1	0	0	1	1
	EFCNI	0	1	0	0	1	1
Reversed	PC	0	2	0	2	3	2
	FCNI	0	0	0	0	0	0
	EFCNI	0	0	0	0	0	0
Spurious	PC	0	3	0	4	3	1
	FCNI	0	3	0	4	3	1
	EFCNI	0	3	0	4	3	1

6.3.1.2 When the events arrive out of order

Table 5 presents the number of erroneous edges in the causal network produced by the three algorithms for varying degree of out-of-order in the event stream. We show the results for the two datasets DS_4 and DS_5 only; the results from the other datasets are consistent with the results from the two datasets.

We make two observations from the results. First, the PC algorithm is more resilient to the out-of-order event arrival than the FCNI or EFCNI algorithm. The number of spurious edges and the number of missing edges are higher in the EFCNI algorithm than in the PC algorithm when the degree of out-of-order is large (i.e., $d_{oo} = 20\%, 25\%$). The reason is that the PC algorithm does not depend on the temporal precedence order for causal network inference at all whereas FCNI and EFCNI do. Second, between the FCNI algorithm and the

EFCNI algorithm, EFCNI results in a comparable number of erroneous edges as PC while FCNI results in a larger number of erroneous edges than EFCNI or PC. The FCNI algorithm completely depends on the temporal order of the events to generate the causal network structure and, consequently, is sensitive to even a small change in the order of the events. In contrast, the EFCNI algorithm employs the OATNI algorithm where the temporal confidence threshold mechanism selects *undirected* edges in the temporal network when the temporal precedence is ambiguous, and this mechanism makes EFCNI more resilient to the changes than FCNI.

Table 5. Number of erroneous edges in an out-of-order event stream for different degrees of out-of-order (d_{oo}) (datasets: DS_4 and DS_5)

Type of Erroneous Edges	Algorithm	d_{oo} for DS_4						d_{oo} for DS_5					
		0 %	5 %	10 %	15 %	20 %	25 %	0 %	5 %	10 %	15 %	20 %	25 %
Missing	PC	0	0	0	0	0	0	1	1	1	1	1	1
	FCNI	0	1	3	4	4	7	1	4	6	7	11	13
	EFCNI	0	0	0	0	1	2	1	1	1	1	1	2
Reversed	PC	2	2	2	2	2	2	3	3	3	3	3	3
	FCNI	0	2	3	5	8	14	0	3	5	6	9	15
	EFCNI	0	0	0	0	0	0	0	0	0	0	0	0
Spurious	PC	4	4	4	4	4	4	3	3	3	3	3	3
	FCNI	4	7	9	12	13	17	3	5	9	11	18	23
	EFCNI	4	4	4	4	4	6	3	3	3	3	4	7

6.3.1.3 When the event stream has changing temporal precedence statistic

Table 6 presents the number of erroneous edges in the causal networks resulting from the three algorithms for the event stream with changing temporal precedence statistic. As expected, the number of erroneous edges from the FCNI or PC algorithm is not affected by these changes because the causal network inference is run every time a new batch of events arrives. In contrast, for the EFCNI algorithm, the number of erroneous edges increases when C_p is lower than δ (at t_3 and t_4). (Note that in such cases the causal network inference is not run.) Additionally, the errors are larger at t_4 than t_3. This is because the higher value of C_p results in a greater difference between the causal network constructed and the true causal network and, more importantly, because the accuracy of the resulting causal network keeps on degrading as we keep on skipping the causal inference. On the other hand, for a batch of events with C_p greater than δ (i.e., at t_1, t_2, t_5, and t_6), the EFCNI algorithm rebuilds the causal network and, consequently, the resulting causal network reflects the true causal network representing the event stream seen thus far. Therefore, at these time points the number of erroneous edges remains the same as if the event stream had no change.

Table 6. Number of erroneous edges in a changing event stream over the six observation points t_1 through t_6 (datasets: DS_4 and DS_5)

Type of Erroneous Edges	Algorithm	DS_4						DS_5					
		t_1	t_2	t_3	t_4	t_5	t_6	t_1	t_2	t_3	t_4	t_5	t_6
Missing	PC	0	0	0	0	0	0	1	1	1	1	1	1
	$FCNI$	0	0	0	0	0	0	1	1	1	1	1	1
	$EFCNI$	0	0	1	2	0	0	1	1	2	4	1	1
Reversed	PC	2	2	2	2	2	2	3	3	3	3	3	3
	$FCNI$	0	0	0	0	0	0	0	0	0	0	0	0
	$EFCNI$	0	0	0	1	0	0	0	0	1	2	0	0
Spurious	PC	4	4	4	4	4	4	3	3	3	3	3	3
	$FCNI$	4	4	4	4	4	4	3	3	3	3	3	3
	$EFCNI$	4	4	5	7	4	4	3	3	5	8	3	3

6.3.2   Comparison of the Running Time of the PC, FCNI, and EFCNI Algorithms

6.3.2.1   When the events arrive in order

Figure 3(a) shows the average running time of the EFCNI, FCNI, and PC algorithms for varying number of event types in the synthetic datasets. In all cases, the running time of the EFCNI algorithm is the shortest while the running time of the PC algorithm is the longest. Clearly, the temporal precedence information helps to reduce the size of condition set and the number of edges for CI tests in both the FCNI and EFCNI algorithms. In addition, the EFCNI algorithm sorts the edges based on their temporal strength and then tests the conditional independence of the weaker edges, which are more likely to fail the tests,

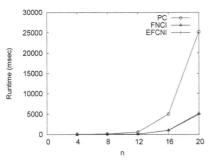

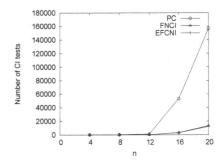

(a) Running time for varying number of event types.

(b) Number of CI tests for varying number of event types.

Fig. 3. Comparison of the running time of the PC, FCNI and EFCNI algorithms for in-order event streams

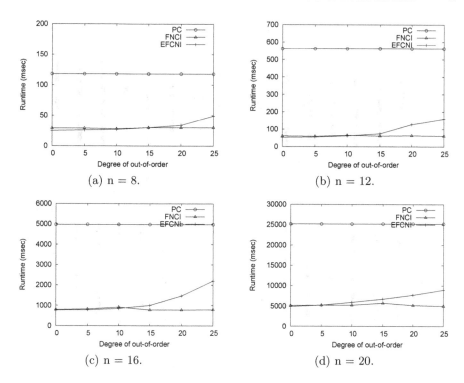

Fig. 4. Comparison of the running time of the PC, FCNI, and EFCNI algorithms for out-of-order event streams

earlier and therefore further reduces the running time. As the number (n) of event types increases, the running time increases in all three algorithms, but the rate of increase is the highest for the PC algorithm and the lowest for the EFCNI algorithm. The same observation is made in the real dataset where the running time of the PC, FCNI, and EFCNI algorithms are 817, 118, and 112 msecs, respectively. These results verify the important role of temporal precedence relationships in reducing the running time.

6.3.2.2   When the events arrive out of order

Figure 4 shows that the EFCNI algorithm performs the fastest causal network inference among the three algorithms when the event stream is in order (i.e., degree of out-of-order $d_{oo} = 0$). As d_{oo} increases, the running time of the EFCNI algorithm increases rapidly. It is due to the strategy that renders the edges with temporal strength lower than θ in the temporal network undirected. Consequently, the number of CI tests increases, resulting in an increase in the running time. On the other hand, as seen in the figure, the running time of FCNI algorithm remains short for the out-of-order event arrivals. However, the FCNI algorithm compromises the accuracy in such an event stream as discussed in Sect. 6.3.1. In addition, the result shows that the running time of the PC algorithm is constant as it is not affected by the out-of-order event arrivals.

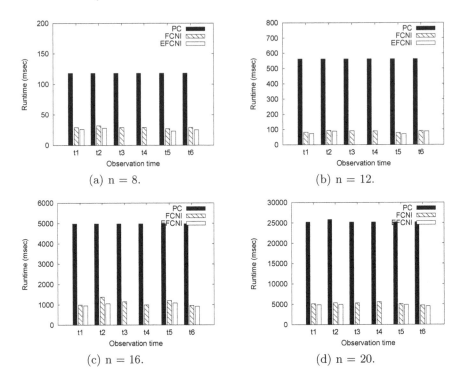

Fig. 5. Comparison of the running time of the PC, FCNI, and EFCNI algorithms for changing event streams

6.3.2.3   When the event stream has changing temporal precedence statistic

Figure 5 shows that the EFCNI algorithm performs the fastest causal network inference among the three algorithms over the event stream with changing temporal precedence statistic. In the figure, for all values of n, the FCNI and PC algorithms perform the CI tests for the causal network inference every time a new batch of events arrives. On the other hand, the EFCNI algorithm performs it only when the precedence statistic changes significantly enough in the event stream. (C_p is greater than δ at t_1, t_2, t_5, and t_6.) The EFCNI algorithm skips the causal inference at t_3 and t_4, which helps to reduce the overall running time.

6.3.3   Comparison of the Number of CI Tests of the PC, FCNI, and EFCNI Algorithms

6.3.3.1   When the events arrive in order

Figure 3 shows that the EFCNI algorithm performs fewer CI tests than the PC and FCNI algorithms in all synthetic datasets. The CI tests are decreased by reducing the size of the condition set and the number of edges to test with the help of the temporal precedence information. In addition, the sorting of the edges (based on the their temporal strengths) helps to reduce the number of CI tests. With an increase in the number of event types (n), the rate of increase in the number of CI tests of the PC algorithm is much higher than that of the EFCNI and FCNI algorithms. A similar observation is made in the real dataset where the number of CI tests of the PC, FCNI, and EFCNI algorithms are 1239, 192, and 176, respectively. These results confirm the important role of temporal precedence relationships in reducing the number of CI tests. Note the result of CI tests (Fig. 3(b)) looks almost the same as that of the running time (Fig. 3(a)). This demonstrates that CI tests are the major performance bottleneck and validates the key idea of our work that reducing the number of CI tests reduces the run time.

6.3.3.2   When the events arrive out of order

Figure 6 shows that, as the degree of out-of-order increases, the number of CI tests of the EFCNI algorithm increases. A higher degree of out-of-order leads to the temporal strengths of more edges lower than the temporal confidence threshold (i.e., θ), and this results in rendering more edges undirected and therefore performing more CI tests. Consequently, as the degree of out-of-order increases, the number of CI tests of the EFCNI algorithm becomes closer to that of the PC algorithm. Note that the PC algorithm is not affected by the out-of-order event arrivals and, thus, the number of CI tests does not change for varying

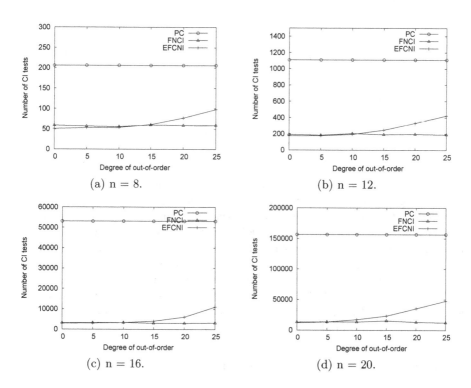

(a) n = 8.

(b) n = 12.

(c) n = 16.

(d) n = 20.

Fig. 6. Comparison of the number of CI tests of the PC, FCNI, and EFCNI algorithms for out-of-order event streams

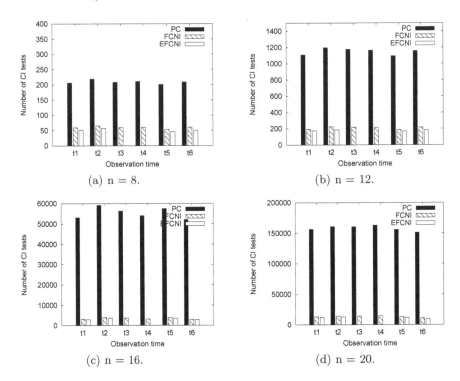

Fig. 7. Comparison of the number of CI tests of the PC, FCNI, and EFCNI algorithms for changing event streams

degree of out-of-order. The results also show that the FCNI algorithm performs the smallest number of CI tests when the events arrive out of order. However, its accuracy is the worst among the three algorithms as discussed in Sect. 6.3.1.

6.3.3.3   When the event stream has changing temporal precedence statistic

Figure 7 shows that the EFCNI algorithm performs the smallest number of CI tests among the three algorithms over the event stream with changing temporal precedence statistic. As discussed earlier, the FCNI and PC algorithms regenerate the causal network with the arrival of every new batch of events while the EFCNI algorithm does it only when the change in the event stream (i.e., C_p) is greater than δ (as seen at t_1, t_2, t_5, and t_6). As a result, the EFCNI algorithm skips causal inference computations involving a large number of CI tests at t_3 and t_4 where the value of C_p is less than δ. As expected, the number of CI tests is highest for the PC algorithm at every observation point t_1 through t_6 due to its highest computational complexity.

6.3.4   Summary of Experiment Results

The EFCNI algorithm is faster than the FCNI algorithm for an event stream where the events arrive in order. The EFCNI algorithm enhances the FCNI algorithm with two additional strategies to reduce the number of CI tests. It has been demonstrated that the CI tests are the performance bottleneck and thus the reduction in the number of CI tests is the key to decreasing the running time of the algorithm. Moreover, unlike the FCNI algorithm which requires events to arrive in order, the EFCNI algorithm can perform the causal network inference accurately even when the events arrive out of order. As the degree of out-of-orderness d_{oo} increases, the accuracy of the EFCNI algorithm comes at the expense of the running time (i.e., the number of CI tests) to some extent.

The EFCNI algorithm is much faster than the PC algorithm in all experiments. In some scenarios (e.g., medical applications like patient health tracking), the accuracy of the result may be more important than the runtime. In case the accuracy of the EFCNI algorithm is not satisfactory in such scenarios – for example, if the event stream has many out-of-order events (i.e., with large d_{oo}) – then the OATNI algorithm can be tweaked to increase the EFCNI algorithm's accuracy by setting θ to a larger value (e.g., toward 100 %). A larger θ value forces the edges more to be undirected and, therefore, makes the EFCNI algorithm behave more like the PC algorithm, thus achieing higher accuracy.

Furthermore, the EFCNI algorithm saves time by avoiding causal inference computations when there is not significant enough changes in the statistic of the event stream.

7 Conclusion and Future Work

In this paper, we presented a novel strategy to exploit temporal precedence relationships to learn the causal network over event streams. First, we introduced the *Order-Aware Temporal Network Inference* algorithm to model temporal precedence information. Then, we presented the *Enhanced Fast Causal Network Inference* algorithm to reduce the running time of learning causal network by reducing the number of performed CI tests significantly. These algorithms efficiently handle the event streams even if the events are out of order and saves the running time further by performing causal inference only if the temporal precedence statistic changes significantly enough. We showed the experiment results to validate our approach by comparing against the state-of-the-art PC and FCNI algorithms.

There are a number of future work in the plan. First, we plan to support cyclic causality. Second, we plan to investigate the effect of concept drift in the causal network inference so that the computations are performed only among the event types affected by the changes. Third, we plan to perform the experiments on datasets with a much larger number (i.e., hundreds to thousands) of event types to show the practicality of our proposed algorithms in a big data environment.

References

1. Barga, R.S., Goldstein, J., Ali, M.H., Hong, M.: Consistent streaming through time: a vision for event stream processing. In: Proceedings of the Third Biennial Conference on Innovative Data Systems Research, CIDR 2007, pp. 363–374 (2007)

2. Zhao, Y., Strom, R.: Exploitng event stream interpretation in publish-subscribe systems. In: Proceedings of the Twentieth Annual ACM Symposium on Principles of Distributed Computing, PODC 2001, pp. 219–228 (2001)
3. Johnson, T., Muthukrishnan, S., Rozenbaum, I.: Monitoring regular expressions on out-of-order streams. In: Proceedings of the IEEE 23rd International Conference on Data Engineering, ICDE 2007, pp. 1315–1319 (2007)
4. Li, M., Liu, M., Ding, L., Rundensteiner, E.A., Mani, M.: Event stream processing with out-of-order data arrival. In: Proceedings of the 27th International Conference on Distributed Computing Systems Workshops, ICDCSW 2007, pp. 67–74. IEEE Computer Society, Washington, DC, USA (2007)
5. Liu, M., Li, M., Golovnya, D., Rundensteiner, E., Claypool, K.: Sequence pattern query processing over out-of-order event streams. In: Proceedings of the IEEE 25th International Conference on Data Engineering, ICDE 2009, pp. 784–795 (2009)
6. Li, J., Tufte, K., Shkapenyuk, V., Papadimos, V., Johnson, T., Maier, D.: Out-of-order processing: a new architecture for high-performance stream systems. Proc. VLDB Endow. **1**(1), 274–288 (2008)
7. Wang, K., Yu, Y.: A query-matching mechanism over out-of-order event stream in iot. Int. J. Ad Hoc Ubiquitous Comput. **13**(3/4), 197–208 (2013)
8. Heckerman, D.: A Bayesian approach to learning causal networks. In: Proceedings of the Eleventh Conference on Uncertainty in Artificial Intelligence, UAI 1995, pp. 285–295 (1995)
9. Ellis, B., Wong, W.H.: Learning causal Bayesian network structures from experimental data. J. Am. Stat. Assoc. **103**(482), 778–789 (2008)
10. Li, G., Leong, T.-Y.: Active learning for causal Bayesian network structure with non-symmetrical entropy. In: Theeramunkong, T., Kijsirikul, B., Cercone, N., Ho, T.-B. (eds.) PAKDD 2009. LNCS, vol. 5476, pp. 290–301. Springer, Heidelberg (2009)
11. Meganck, S., Leray, P., Manderick, B.: Learning causal Bayesian networks from observations and experiments: a decision theoretic approach. In: Torra, V., Narukawa, Y., Valls, A., Domingo-Ferrer, J. (eds.) MDAI 2006. LNCS (LNAI), vol. 3885, pp. 58–69. Springer, Heidelberg (2006)
12. Pearl, J.: Causality: Models, Reasoning and Inference, 2nd edn. Cambridge University Press, Cambridge (2009)
13. Spirtes, P., Glymour, C.N., Scheines, R.: Causality from probability. In: Proceedings of the Conference on Advanced Computing for the Social Sciences, ACSS 1990 (1990)
14. Spirtes, P., Glymour, C., Scheines, R.: Causation, Prediction, and Search. MIT Press, Cambridge (2000)
15. Cheng, J., Greiner, R., Kelly, J., Bell, D., Liu, W.: Learning Bayesian networks from data: an information-theory based approach. Artif. Intell. **137**(1–2), 43–90 (2002)
16. Pearl, J.: Causal diagrams for empirical research. Biometrika **82**, 669–710 (1995)
17. Spirtes, P., Meek, C.: Learning Bayesian networks with discrete variables from data. In: Proceedings of the First International Conference on Knowledge Discovery and Data Mining, KDD 1995, pp. 294–299 (1995)
18. Chow, Y.S., Teicher, H.: Probability Theory: Independence, Interchangeability, Martingales. Springer, New York (1978)
19. Prakasa Rao, B.: Conditional independence, conditional mixing and conditional association. Ann. Inst. Stat. Math. **61**(2), 441–460 (2009)
20. Popper, K.: The Logic of Scientific Discovery, Reprint edition. Routledge, New York (1992)

21. Hamilton, H.J., Karimi, K.: The TIMERS II algorithm for the discovery of causality. In: Ho, T.-B., Cheung, D., Liu, H. (eds.) PAKDD 2005. LNCS (LNAI), vol. 3518, pp. 744–750. Springer, Heidelberg (2005)

22. Utrera, A.C., Olmedo, M.G., Callejon, S.M.: A score based ranking of the edges for the pc algorithm. In: Proceedings of the 4th European Workshop on Probabilistic Graphical Models, PGM 2008, pp. 41–48 (2008)

23. Acharya, S., Lee, B.S.: Fast causal network inference over event streams. In: Bellatreche, L., Mohania, M.K. (eds.) DaWaK 2013. LNCS, vol. 8057, pp. 222–235. Springer, Heidelberg (2013)

24. Chickering, D.M.: Learning equivalence classes of Bayesian-network structures. J. Mach. Learn. Res. **2**, 445–498 (2002)

25. Borchani, H., Chaouachi, M., Ben Amor, N.: Learning causal Bayesian networks from incomplete observational data and interventions. In: Mellouli, K. (ed.) ECSQARU 2007. LNCS (LNAI), vol. 4724, pp. 17–29. Springer, Heidelberg (2007)

26. de Campos, L.M.: A scoring function for learning Bayesian networks based on mutual information and conditional independence tests. J. Mach. Learn. Res. **7**, 2149–2187 (2006)

27. Bishop, Y.M., Fienberg, S.E., Holland, P.W.: Discrete Multivariate Analysis: Theory and Practice. MIT Press, Cambridge (1975)

28. Kullback, S.: Information Theory and Statistics, 2nd edn. Dover Publication, New York (1968)

29. Verma, T., Pearl, J.: Causal networks: Semantics and expressiveness. In: Proceedings of the Fourth Annual Conference on Uncertainty in Artificial Intelligence, UAI 1988, pp. 69–78 (1988)

30. Geiger, D., Pearl, J.: On the logic of causal models. In: Proceedings of the Fourth Annual Conference on Uncertainty in Artificial Intelligence, UAI 1988, pp. 3–14 (1988)

31. Pearl, J.: Graphs, causality, and structural equation models. Sociol. Methods Res. **27**, 226–284 (1998)

32. Tsamardinos, I., Brown, L.E., Aliferis, C.F.: The max-min hill-climbing Bayesian network structure learning algorithm. Mach. Learn. **65**(1), 31–78 (2006)

33. Frank, A., Asuncion, A.: UCI machine learning repository (2010). http://archive.ics.uci.edu/ml/datasets/Diabetes

Learning Through Non-linearly Supervised Dimensionality Reduction

Josif Grabocka$^{(\boxtimes)}$ and Lars Schmidt-Thieme

Information Systems and Machine Learning Lab, Samelsonplatz 22,
31141 Hildesheim, Germany
{josif,schmidt-thieme}@ismll.uni-hildesheim.de

Abstract. Dimensionality reduction is a crucial ingredient of machine learning and data mining, boosting classification accuracy through the isolation of patterns via omission of noise. Nevertheless, recent studies have shown that dimensionality reduction can benefit from label information, via a joint estimation of predictors and target variables from a lowf-rank representation. In the light of such inspiration, we propose a novel dimensionality reduction which simultaneously reconstructs the predictors using matrix factorization and estimates the target variable via a dual-form maximum margin classifier from the latent space. Compared to existing studies which conduct the decomposition via linearly supervision of targets, our method reconstructs the labels using nonlinear functions. If the hyper-plane separating the class regions in the original data space is non-linear, then a nonlinear dimensionality reduction helps improving the generalization over the test instances. The joint optimization function is learned through a coordinate descent algorithm via stochastic updates. Empirical results demonstrate the superiority of the proposed method compared to both classification in the original space (no reduction), classification after unsupervised reduction, and classification using linearly supervised projection.

Keywords: Machine learning · Dimensionality reduction · Feature extraction · Matrix factorization · Supervised dimensionality reduction

1 Introduction

Dimensionality reduction is an important ingredient of machine learning and data mining. The benefits of projecting data to latent spaces constitute in (i) converting large dimensionality datasets into feasible dimensions, but also (ii) improving the classification accuracy of small and medium datasets [1]. The scope of this work lies on improving prediction accuracy rather than ensuring scalability. There exists a trade-off between accurate and scalable methods, concretely a plain unsupervised dimensionality reduction is often advised for scalability (fast classification) purposes [2]. Via carefully tuned dimensionality reduction (aka feature extraction) we are able to retrieve the necessary patterns from the datasets, by leaving out the noise. Traditional dimensionality reduction

© Springer-Verlag Berlin Heidelberg 2015
A. Hameurlain et al. (Eds.): TLDKS XVII, LNCS 8970, pp. 74–96, 2015.
DOI: 10.1007/978-3-662-46335-2_4

(described in Sect. 2.1) has been focused on extracting features prior to classification. Such a mentality has been recently found to perform non-optimal [3,4], since the features are not directly extracted/optimized for boosting classification. This discrepancy is created because the objective function (loss) of the unsupervised decomposition process is different from the one used during the evaluation of the prediction accuracy. Typically the L2 (Euclidean) error is used in approximating the original predictor values from the low-rank data, while a logistic, or hinge loss function, for the classification of targets. In order to solve this challenge, there have been attempts to incorporate class supervision into feature extraction, (mentioned in Sect. 2.3), such that the latent features are guided to enforce the discernment/separation of instances belonging to opposite classes in the reduced space.

Throughout this work we propose a principle, (details in Sect. 3.1), according to which dimensionality reduction should optimize the latent features through the same optimization function as the final classification method, thereby ensuring that the classification accuracy in the latent space is optimized. Inspired by the accuracy success of Support Vector Machines (SVM) which is largely credited to the kernel trick approach, we propose a novel supervised dimensionality reduction that incorporates kernel-based classification in the reduced dimension (Sect. 3). The novelty relies on defining a joint dimensionality reduction via matrix factorization, in parallel to a dual-form kernel-based maximum margin classification in the latent space. The reduced data is simultaneously updated in a coordinate descent fashion in order to optimize both loss terms. Experimental results (Sect. 4) demonstrate the superiority of the proposed method compared to both unsupervised dimensionality reduction and classification in the original space. The main contribution of this work are:

1. Defined a supervised dimensionality reduction with a kernel-based target variable estimation
2. Reviewed and elaborated the state of the art in supervised dimensionality reduction
3. Derived a coordinate descent algorithm which simultaneously learns the latent factors for the reconstruction of predictors and the accuracy over target
4. Compared the paradigms of linearly versus non-linearly supervised dimensionality reduction
5. Provided empirical results to demonstrate the superiority of the proposed method

2   Related Work

2.1   Dimensionality Reduction

Dimensionality reduction is a field of computer science that focuses on extracting lower dimensionality features from datasets [1]. Numerous techniques exist for extracting features. Principal Component Analysis (PCA) is a famous approach involving orthogonal transformations and selecting the topmost principal components, which preserve necessary variance [5]. Alternatively, Singular Value Decomposition decomposes a dataset into latent unitary, nonnegative diagonal and conjugate transpose unitary matrices [1].

Further elaborations of dimensionality reductions involve nonlinear decomposition of data [6]. For instance kernel PCA replaces the linear operations of PCA through nonlinear mappings in a Reproducing Kernel Hilbert Space [7]. The whole subfield of manifold learning elaborates, as well, on nonlinear projections. Specifically, Sammon's mappings preserves the structure of instance distances in the reduced space [8], while principal curves embed manifolds using standard geometric projections [9]. More nonlinear dimensionality algorithms are described in [10]. In addition, temporal dimensionality reduction have been proposed in scenarios where the time difference of observations is not evenly spaced [11]. The field of Gaussian Processes have been extended to dimensionality reduction through the Gaussian Processes Latent Variable Models (GPLVM) [12,13]. In comparison to PCA, the GPLVM models define nonlinear approximative functions for the predictor values [12].

2.2   Matrix Factorization

Matrix factorization refers to a family of decompositions which approximate a dataset as a product of latent matrices of typically lower dimensions. A generalization and categorization of the various proposed factorization models is elaborated in [14], where factorizations are seen as applications of the Bregman Divergence paradigm. The learning of the decomposition is typically conducted by defining a L2-norm and updating the latent matrices via a stochastic gradient descent algorithm [15]. Matrix factorization has been applied in a range of domains, ranging from recommender systems where decomposition focuses on collaborative filtering of sparse user-item ratings [16], up to time series dimensionality reduction [17]. The Matrix Factorization approach is a special instance of a probabilistic PCA, while it extends the functionality of a PCA by adding bias terms [15]. In terms of similarities, a factorization can be also characterized as a biased probabilistic SVD. In a broader sense, the linear approximation of predictors can be also interpreted as an instance of the GPLVM models for a linear (polynomial of degree one) kernel [12].

2.3   Supervised Dimensionality Reduction

In addition to the standard dimensionality reduction and Matrix Factorization, there has been attempts to utilize the labels information, therefore dictating a supervised projection. Fisher's linear discriminant analysis is a popular supervised projection method [18]. The classification accuracy loss objective functions occurring in literature vary from label least square regression [19], to generalized linear models [20], linear logistic regression [3], up to hinge loss [4,21]. Another study aimed at describing the target variable as being conditionally dependent on the features [22]. Other families of supervisions strive for preserving the neighborhood structure of intra-class instances [23], or links in a semi supervised scenarios [24]. The GPLVM models have been, as well, adopted for discriminative classification [25]. A self-contained description of the state of the art in linearly supervised dimensionality reduction is offered in Sect. 3.4. In comparison to the aforementioned methods, we propose a supervised dimensionality reduction with a kernel-based classifier that directly optimizes the dual formulation in the projected space.

3   Proposed Method

3.1   Principle

The method proposed in this study relies on the principle that feature extraction, analogously referred also as dimensionality reduction, should not be conducted "ad-hoc" or via particular heuristics. Most of the classification tasks have a unifying objective, which is to improve classification accuracy. In that context we are referring as "ad-hoc" to the family of feature extraction techniques that don't directly optimize their loss functions for classification accuracy. Unsupervised projection is not optimized for the same loss function which is used during the evaluation of the target variable. Techniques such as SVD or Matrix Factorization focus solely on approximating the predictors of the original data. Unfortunately, unsupervised decompositions pose the risk of losing the signal relevant to the prediction accuracy. While such approaches approximate the original data, they become vulnerable to the noise present in the observed predictors' values. Stated else-wise, we believe that instance labels should guide the feature extraction, such that the utilization of the extracted features improves accuracy. In that perspective, we propose a feature extraction method which operates by optimizing a joint objective function composed of the feature extraction term and also the classification accuracy term. In comparison with similar feature extraction ideas reviewed in Sect. 2.3, which use linear classifiers in the optimization, we propose a novel method which learns a nonlinear SVM over the projected space via jointly optimizing a dual form together with dimensionality reduction. Further details will be covered throughout Sect. 3 which is organized progressively. Initially the unsupervised dimensionality reduction is explained and then the state of the art in linearly supervised decomposition. Finally the stage is ready for introducing our novel method on non-linearly supervised dimensionality reduction.

3.2   Introduction to Supervised Dimensionality Reduction

Unsupervised dimensionality reduction (e.g.: matrix factorization described in Sect. 3.3), is guided only by the reconstruction loss. Such an approach does not take into consideration the classification accuracy impact of the extracted features, therefore the produced reduced dimensionality data is not optimized to improve accuracy. In order to overcome such a drawback, the so called supervised dimensionality reduction has been proposed by various authors (see Sect. 2.3).

The key commonalities of those supervised dimensionality methods rely on defining a joint optimization function, consisting of the reconstruction loss terms and the classification accuracy terms.

The typical classification accuracy loss term focuses on defining a classifier in the latent space, i.e. $U \in \mathbb{R}^{(n+n')\times d}$, via a hyperplane defined by the weights vector $W \in \mathbb{R}^d$, such that the weights can correctly classify the training instances of U in order to match observed label $Y \in \mathbb{R}^n$. Equation 1 defines a cumulative joint optimization function using a reconstruction term for the predictors, denoted $F_R(X, U, V)$, and a classification accuracy term, denoted $F_{CA}(Y, U, W)$. The trick of such a joint optimization constitutes on updating the low-rank data U simultaneously, in order to minimize both F_R and F_{CA} via gradient descent on both loss terms. The hyper parameter β is a switch which balances the impact of reconstruction vs classification accuracy. Throughout this paper we evaluate the binary classification problem, even though the explained methods could be trivially transferred to multi-nominal target variables by employing the one-vs-all technique. Should that be needed, we would have to build as many classifiers as there are categories in the target variable, while each classifier would treat one category value as the positive class and all the remaining categories as the negative class. In addition to the reconstruction F_R and the classification accuracy F_{Ca} loss terms, the model has additive regularization terms parametrized by coefficients $\lambda_U, \lambda_V, \lambda_W$. Such a regularization helps the model avoid overfitting and enables a better generalization over the test instances.

$$F(X, Y, U, V, W) = \beta\, F_R(X, U, V) + (1 - \beta)\, F_{CA}(Y, U, W) \qquad (1)$$

3.3   Matrix Factorization as Dimensionality Reduction

Matrix factorization is a dimensionality reduction technique which decomposes a dataset $X \in \mathbb{R}^{(n+n')\times m}$ matrix of n training instances and n' testing instances, per m features, into two smaller matrices of dimensions $U \in \mathbb{R}^{(n+n')\times d}$ and $V \in \mathbb{R}^{d\times m}$ [15]. The latent/reduced projection of the original data X is the latent matrix U, where d is the dimensionality of the projected space. Typically d is much smaller than m, meaning that the dimensionality is reduced. In case $d < m$, then U is nominated as the low-rank representation of X. Otherwise, if $d > m$ a *non-grata* inflation phenomenon is achieved. Such decomposition is expressed in a form of a regularized reconstruction loss, denoted $F_R(X, U, V)$ and depicted in Eq. 2. The optimization of such a function aims at computing latent matrices U, V such that their dot product approximates the original matrix X via an Euclidean distance ($L2$ norm) loss. In addition to the $L2$ reconstruction norm, we also add $L2$ regularization terms weighted by factors λ_U, λ_V in order to avoid over-fitting.

$$\underset{U,V}{\text{argmin}}\ \ F_R(X, U, V) = ||X - UV||^2 + \lambda_U ||U||^2 + \lambda_V ||V||^2 \qquad (2)$$

Bias terms, $B_U \in \mathbb{R}^{(n+n')\times 1}, B_V \in \mathbb{R}^{1\times m}$ are added to the reconstruction loss [15], such that each element of B_U incorporates the prior belief value of the respective instance, while each element of B_V the prior belief value of the respective feature. More concretely the loss can be expanded as a reconstruction of each cell $X_{i,j}$ as depicted by Eq. 3.

$$\underset{U,V,B_U,B_V}{\mathrm{argmin}} \quad F_R(X,U,V) = \sum_{i=1}^{n+n'} \sum_{j=1}^{m} \left(X_{i,j} - \left(\sum_{k=1}^{d} U_{i,k} V_{k,j} + B_{U_i} + B_{V_j} \right) \right)^2$$
$$+ \lambda_U \sum_{i=1}^{n+n'} \sum_{k=1}^{d} U_{i,k}^2 + \lambda_V \sum_{k=1}^{d} \sum_{j=1}^{m} V_{k,j}^2 \tag{3}$$

In order to learn the Matrix Factorization defined in Eq. 3 we need to define the gradients to be used for updating our latent matrices. Stochastic Gradient Descent is a fast optimization technique for factorizations [15] and operates by reducing the approximation error of each cell (i,j) of X. Therefore, we can represent the reconstruction loss F_R as sum of smaller loss terms $F_{Ri,j}$, per each cell (i,j) of the original dataset X. Such a decomposition will later enable the stochastic gradient descent to optimize for each small loss term stochastically, i.e. the indices (i,j) will be visited randomly.

$$F_R(X,U,V) = \sum_{i=1}^{n+n'} \sum_{j=1}^{m} F_R(X,U,V)_{i,j} \tag{4}$$

$$F_R(X,U,V)_{i,j} = \beta \left(X_{i,j} - \left(\sum_{k=1}^{d} U_{i,k} V_{k,j} + B_{U_i} + B_{V_j} \right) \right)^2$$
$$+ \lambda_U \frac{1}{m} \sum_{k=1}^{d} U_{i,k}^2 + \lambda_V \frac{1}{n+n'} \sum_{k=1}^{d} V_{k,j}^2 \tag{5}$$

The gradients of the latent data U, V with respect to the reconstruction loss are computed as the first derivative of the loss. The error in approximating a cell $X_{i,j}$ is defined as $e_{i,j}$ and can be pre-computed for scalability. As can be observed from the gradients of Eqs. 6–9, the pre-computed error term $e_{i,j}$ is used in all gradients.

$$e_{i,j} = X_{i,j} - \sum_{k=1}^{d} U_{i,k} V_{k,j} - B_{U_i} - B_{V_j} \tag{6}$$

$$\frac{\partial F_R(X,U,V)_{i,j}}{\partial U_{i,k}} = -2\beta\, e_{i,j}\, V_{k,j} + 2\lambda_U \frac{1}{m} U_{i,k} \tag{7}$$

$$\frac{\partial F_R(X,U,V)_{i,j}}{\partial V_{k,j}} = -2\beta\, e_{i,j}\, U_{i,k} + 2\lambda_V \frac{1}{n+n'} V_{k,j} \tag{8}$$

$$\frac{\partial F_R(X,U,V)_{i,j}}{\partial B_{U_i}} = \frac{\partial F_R(X,U,V)_{i,j}}{\partial B_{V_j}} = -2\beta\, e_{i,j} \tag{9}$$

3.4 Linearly Supervised Dimensionality Reduction

The linear supervision of the dimensionality reduction refers to the inclusion of a linear classification loss term to the objective function, expressed as F_{CA} in Eq. 1. The addition of the linear classification loss term enforces the instances of different classes to be linearly separable in the low-rank space. Various loss terms have been proposed depending on the utilized linear classifier. Before explaining the different losses, we introduce the predicted value of instance i as $\hat{Y}_i$ and defined in Eq. 10. The predicted value is the dot product of the instance values $U_{i,:} \in \mathbb{R}^d$ and linear weights $W \in \mathbb{R}^d$. In addition, the bias term for the instance $B_{U_i} \in \mathbb{R}$ and the bias of the classification weight vector $W_0 \in \mathbb{R}$ are summed up.

$$\hat{Y}_i = B_{U_i} + W_0 + \sum_{k=1}^{d} U_{i,k} W_k, \quad \forall i \in \mathbb{N}_i^n \tag{10}$$

Loss terms quantify the degree of violation that a classifier exhibits from the desired (perfect) prediction accuracy. Concretely the least square loss measures the L2 distance between the true targets Y and predicted vales $\hat{Y}$. In the context of linearly supervised reduction [19], the least-squares loss term can be defined as shown in Eq. 11. Similar to the regression case, least squares is adopted for classification by treating the target values as $Y \in \{-1, 1\}^n$, while predicted positive values $\hat{Y}$ indicate a positive class and vice versa.

$$F_{CA}(Y, U, W)^{LS} = \sum_{i=1}^{n} \left(Y_i - \hat{Y}_i \right)^2 + Reg(U, W), \quad \forall i \in \mathbb{N}_i^n \tag{11}$$

The logistic loss has been applied to guide the decomposition by minimizing the target prediction error along a sigmoid curve [3]. Equation 12 presents the loss, while the target values are expected to be in the range $Y \in \{0, 1\}^n$. Please note that the sigmoid function is defined as: $\text{sigmoid}(\hat{Y}) = \frac{1}{1 + e^{-\hat{Y}}}$.

$$F_{CA}(Y, U, W)^{LO} = \sum_{i=1}^{n} -Y_i \log(\text{sigmoid}(\hat{Y}_i)) - (1 - Y_i)$$
$$\times \log\left(1 - \text{sigmoid}(\hat{Y}_i)\right) + Reg(U, W) \tag{12}$$

Another strong linear classifier is the hinge loss, which represents the underlying foundation of the Support Vector Machines is depicted in Eq. 13. The hinge loss has been also applied to supervised reduction [4]. The hinge loss is also called a maximum margin loss because it tries to find a margin of unit size between the hyperplane W and the region of each class.

$$F_{CA}(Y, U, W)^{HI} = \sum_{i=1}^{n} \max(0, 1 - Y_i \hat{Y}_i) + Reg(U, W), \quad \forall i \in \mathbb{N}_i^n \tag{13}$$

Unfortunately the hinge loss is not differentiable at $Y\hat{Y} = 1$, therefore a smoothed variant of the hinge loss [21] is preferred in cases where a gradient based optimization is needed (Eq. 14).

$$F_{CA}(Y, U, W)^{SH} = \sum_{i=1}^{n} \left(\begin{cases} 1 - Y_i\hat{Y}_i & Y_i\hat{Y}_i < 0 \\ \frac{1}{2}\left(1 - Y_i\hat{Y}_i\right)^2 & 0 \leq Y_i\hat{Y}_i < 1 \\ 0 & Y_i\hat{Y}_i \geq 1 \end{cases} \right) + Reg(U, W) \quad (14)$$

The regularization term is a L2 norm and defined in Eq. 15. The regularization parameters λ_U, λ_W control the complexity of the model and avoid overfitting.

$$Reg(U, W) = \sum_{i=1}^{n+n'} \sum_{k=1}^{d} U_{i,k}^2 + \sum_{k=1}^{d} W_k^2 \quad (15)$$

The Advantage of Supervised Decomposition relies on using the label information to guide the projection. In that way, any noise which is present in the observed data can be eliminated in the low-rank representation.

In order to show the advantage of the supervised decomposition, we present the experiment of Fig. 1. A 2-dimensional synthetic dataset of ten instances, belonging to two classes (red, blue) is depicted in sub-figure (a). Please note that the original data are **linearly separable** by a hyperplane. Then, we added a random variable X_3 (shown in (b)) of uniform random values between $[-1, 1]$. The experiment aims at reducing the 3-dimensional noisy data back to 2-dimensions using both unsupervised and supervised dimensionality reductions. As can be observed, the unsupervised projection is affected by the added noise and the resulting 2-dimensional data in (c) is not anymore linearly separable. In contrast, the linearly supervised decomposition can benefit from a linear classification accuracy loss term to separate instances by label. A smooth hinge loss supervised decomposition was applied to the decomposition of (d). Please note that the resulting 2-dimensional projection depicted in (d) is linearly separable as the original data. The experiment demonstrates that a supervised decomposition has stronger immunity towards the presence of noise in the data. For the sake of reproducibility, the parameters used during the experiment are provided in the caption note.

Learning the Linearly Supervised Decomposition is carried on through optimizing the latent weights U and W by taking a step in the first derivative of the classification accuracy term F_{CA}. In comparison to full gradient approaches, stochastic techniques operate by eliminating the error of a random single instance i, i.e. optimizing for F_{CAi}. Since the gradient computations for each instance are much simpler than for the full dataset, the stochastic gradient descent computes faster than the full gradient learning.

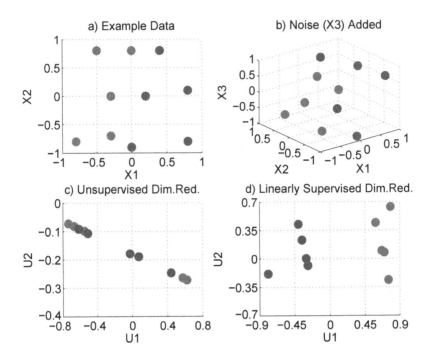

Fig. 1. Supervised reduction: (a) original data with two classes (blue and red); (b) random noise variable (X3) added; (c) unsupervised dimensionality reduction through matrix factorization (from (b) to (c)); (d) linearly supervised maximum margin dimensionality reduction (from (b) to (d)). Parameters: $\eta_R = \eta_{CA} = 0.001$, $\lambda_U = \lambda_V = 0.0001$, $\lambda_V =$ and $\beta = 0.4$ (Color figure online).

More specifically, the gradients of the least-squares loss term are shown in Eqs. 16–18 and are the result of the first derivative with respect to each cell of U, W and the biases B_{U_i}, W_0.

$$\frac{\partial F_{CA}(Y, U, W)_i^{LS}}{\partial U_{i,k}} = -2\left(Y_i - \hat{Y}_i\right)W_k + 2\frac{\lambda_U}{m}U_{i,k} \tag{16}$$

$$\frac{\partial F_{CA}(Y, U, W)_i^{LS}}{\partial W_k} = -2\left(Y_i - \hat{Y}_i\right)U_{i,k} + 2\frac{\lambda_W}{n}W_k \tag{17}$$

$$\frac{\partial F_{CA}(Y, U, W)_i^{LS}}{\partial B_{U_i}} = \frac{\partial F_{CA}(Y, U, W)_i^{LS}}{\partial W_0} = -2\left(Y_i - \hat{Y}_i\right) \tag{18}$$

The logistic loss has derivatives similar to the least square loss with the difference being the inclusion of the sigmoid of the predicted target value $\hat{Y}$. The detailed gradients for the latent weights, with respect to the logistic loss term $F_{CA}(Y, U, W)^{LO}$ are presented in Eqs. 19–21.

$$\frac{\partial F_{CA}(Y,U,W)_i^{LO}}{\partial U_{i,k}} = -2\left(Y_i - \text{sigmoid}(\hat{Y}_i)\right)W_k + 2\frac{\lambda_U}{m}U_{i,k} \tag{19}$$

$$\frac{\partial F_{CA}(Y,U,W)_i^{LO}}{\partial W_k} = -2\left(Y_i - \text{sigmoid}(\hat{Y}_i)\right)U_{i,k} + 2\frac{\lambda_W}{n}W_k \tag{20}$$

$$\frac{\partial F_{CA}(Y,U,W)_i^{LO}}{\partial B_{U_i}} = \frac{\partial F_{CA}(Y,U,W)_i^{LS}}{\partial W_0} = -2\left(Y_i - \text{sigmoid}(\hat{Y}_i)\right) \tag{21}$$

The optimization of the smooth hinge loss requires the optimization of the three conditional steps present in the loss function, for the regions $Y\hat{Y} < 0$, $0 \le Y\hat{Y} < 1$ and $Y\hat{Y} < 1$. The computation of the gradients follows a similar practice as the other aforementioned loss types. First derivatives of the smooth hinge loss term F_{CA}^{SH} are computed per each latent weight U, W, B_U, W_0. The derived gradients are shown in Eqs. 22–24.

$$\frac{\partial F_{CA}(Y,U,W)_i^{SH}}{\partial U_{i,k}} = \left(\begin{cases} -Y_i\hat{Y}_iW_k & Y_i\hat{Y}_i < 0 \\ -\left(1 - Y_i\hat{Y}_i\right)W_k & 0 \le Y_i\hat{Y}_i < 1 \\ 0 & Y_i\hat{Y}_i \ge 1 \end{cases}\right) + 2\frac{\lambda_U}{m}U_{i,k} \tag{22}$$

$$\frac{\partial F_{CA}(Y,U,W)_i^{SH}}{\partial W_k} = \left(\begin{cases} -Y_i\hat{Y}_iU_{i,k} & Y_i\hat{Y}_i < 0 \\ -\left(1 - Y_i\hat{Y}_i\right)U_{i,k} & 0 \le Y_i\hat{Y}_i < 1 \\ 0 & Y_i\hat{Y}_i \ge 1 \end{cases}\right) + 2\frac{\lambda_W}{n}W_k \tag{23}$$

$$\frac{\partial F_{CA}(Y,U,W)_i^{SH}}{\partial B_{U_i}} = \frac{\partial F_{CA}(Y,U,W)_i^{SH}}{\partial W_0} = \begin{cases} -Y_i\hat{Y}_i & Y_i\hat{Y}_i < 0 \\ -\left(1 - Y_i\hat{Y}_i\right) & 0 \le Y_i\hat{Y}_i < 1 \\ 0 & Y_i\hat{Y}_i \ge 1 \end{cases} \tag{24}$$

A Final Learning Algorithm is constructed by applying the defined gradients in a stochastic gradient descent approach over the reconstruction and accuracy loss terms. Algorithm 1 concatenates all the pieces of the learning process. The learning process is separated into two main sections, namely (i) the updates with respect to the reconstruction loss and (ii) the updates with respect to the classification accuracy loss terms. The first loop iterates over all cells of X indexed by row-column pairs (i,j), and also updates all the cells of U according to the error present in approximating $X_{i,j}$. Similarly, the second loop iterates over the train targets Y_i and corrects the classification errors. The name of the loss term (LT) is a generic placeholder and aforementioned gradients of each loss (least-squares, logistic and hinge) can be directly plugged in.

Updates are applied to all the cells of U, V, W and the biases B_U, B_V, W_0 in a stochastic gradient fashion, i.e. visited randomly. The random updates speed up the learning process because the continuous update of columns from a single row is avoided. An update relies on decrementing the value of a cell in the direction of the aforementioned gradients. The magnitude of the decrement step is controlled using a learning rate parameters. Technically, there are two learning

Algorithm 1. Learning Algorithm: Linearly Supervised Dim. Red.

Input: Dataset matrix $X \in \mathbb{R}^{(n+n')\times m}$, Labels vector $Y \in \mathbb{R}^n$, **Parameters:** {Optimization switch β, Latent dimensions d, Learning rates η_R, η_{CA}, Regularizations $\lambda_U, \lambda_V, \lambda_W$ }

Output: U, V, B_U, B_V, W, W_0

Initialize randomly $U \in \mathbb{R}^{(n+n')\times d}$, $V \in \mathbb{R}^{d\times m}$, $W \in \mathbb{R}^d$, $B_U \in \mathbb{R}^{(n+n')\times 1}$, $B_V \in \mathbb{R}^{1\times m}$, $W_0 \in \mathbb{R}$

while $F_R + F_{CA}$ **not** reached an optimum **do**

  for $\forall (i,j,k) \in (\{1...(n+n')\}, \{1...m\}, \{1...d\})$ *in random order* **do**

$$U_{i,k} \leftarrow U_{i,k} - \eta_R \frac{\partial F_R(X,U,V)_{i,j}}{\partial U_{i,k}}$$
$$V_{k,j} \leftarrow V_{k,j} - \eta_R \frac{\partial F_R(X,U,V)_{i,j}}{\partial V_{k,j}}$$
$$B_{U_i} \leftarrow B_{U_i} - \eta_R \frac{\partial F_R(X,U,V)_{i,j}}{\partial B_{U_i}}$$
$$B_{V_j} \leftarrow B_{V_j} - \eta_R \frac{\partial F_R(X,U,V)_{i,j}}{\partial B_{V_j}}$$

  end for

  $LT \leftarrow \{LS, LO, SH\}$ {LT stands for 'Loss Type'}

  for $\forall (i,k) \in (\{1...n\}, \{1...d\})$ *in random order* **do**

$$U_{i,k} \leftarrow U_{i,k} - \eta_{CA} \frac{\partial F_{CA}(Y,U,W)_i^{LT}}{\partial U_{i,k}}$$
$$W_k \leftarrow W_k - \eta_{CA} \frac{\partial F_{CA}(Y,U,W)_i^{LT}}{\partial W_k}$$
$$B_{U_i} \leftarrow B_{U_i} - \eta_{CA} \frac{\partial F_{CA}(Y,U,W)_i^{LT}}{\partial B_{U_i}}$$
$$W_0 \leftarrow W_0 - \eta_{CA} \frac{\partial F_{CA}(Y,U,W)_i^{LT}}{\partial W_0}$$

  end for

end while

return U, V, B_U, B_V, W, W_0

rates for each relation η_R and η_{CA}, which can also have equal values. The second cycle iterates over all the low-rank instances $U_{i,:}$ and the classification weights vector W_k.

$$\hat{Y}_t^{LS,SH} = \text{sign}(W_0 + B_{U_t} + \sum_{k=1}^d U_{t,k} W_k), \qquad t \in \mathbb{N}_{n+1}^{n'} \quad (25)$$

$$\hat{Y}_t^{LO} = \text{sigmoid}(W_0 + B_{U_t} + \sum_{k=1}^d U_{t,k} W_k) > \frac{1}{2} ? 1 : 0, \quad t \in \mathbb{N}_{n+1}^{n'} \quad (26)$$

Once the weights are learned, then the prediction of the test instances can be produced as shown in Eqs. 25–26. For the least squares and the smooth hinge loss the sign function defines the prediction, i.e.: the positive class for positive predictive values, otherwise the negative class. On the other side, the logistic loss is defined in Eq. 26 as one of $\{0, 1\}$ based on the value threshold of 0.5.

3.5 Nonlinearly Supervised Dimensionality Reduction

In comparison to previous approaches that propose linear models, in this study we propose a kernel-based **binary** classifier approach in the latent space U.

Let us initially define the classification accuracy loss term, denoted $F_{CA}(Y, U, W)$, in Eq. 27, in form of a maximum margin soft SVMs with hinge loss [26]. Such form of the SVMs is called the primal form. The parameter C scales the penalization of the instances violating the distances from the maximum margin. Please note that W_0 is the intercept bias term of the hyperplane weights vector W.

$$\underset{U,W}{\operatorname{argmin}}\ F_{CA}(Y, U, W) = \frac{1}{2}||W||^2 + C \sum_{i=1}^{n} \xi_i \tag{27}$$

$$\text{s.t:}\ Y_i(\langle W, U_i \rangle + B_{U_i} + W_0) \geq 1 - \xi_i, \quad i = 1, ..., n$$

$$\xi_i \geq 0, \quad i = 1, ..., n$$

Unfortunately the primal form doesn't support kernels, therefore we have to convert the optimization functions into the dual form Eq. 28. In order to get rid of the inequality constraint we apply Lagrange multipliers to include the inequalities by introducing dual variables α_i per instance and adding $\alpha_i(y_i(\langle W, U_i \rangle + W_0))$ to the optimization function for all instance i. Then we solve the objective function for W and W_0 by equating the first derivative to zero. Putting the derived expressions of W and W_0 to the objective function, we obtain the so-called dual representation optimization:

$$\underset{U,\alpha}{\operatorname{argmin}}\ F_{CA}(Y, U, \alpha) = \frac{1}{2} \sum_{i=1}^{n} \sum_{l=1}^{n} \alpha_i \alpha_l Y_i Y_l \langle [U_{i,*}, B_{U_i}], [U_{t,*}, B_{U_l}] \rangle - \sum_{i=1}^{n} \alpha_i \tag{28}$$

$$\text{s.t:}\ 0 \leq \alpha_i \leq C, \quad i = 1, ..., n; \quad \text{and} \quad \sum_{i=1}^{n} \alpha_i Y_i = 0$$

Once the optimization model is build any new test instance U_t can be classified in terms of learned α as shown in Eq. 29.

$$\hat{Y}_t = \operatorname{sign}\left(\sum_{i=1}^{n} \alpha_i Y_i \langle [U_{i,*}, B_{U_i}], [U_{t,*}, B_{U_l}] \rangle + W_0 \right) \tag{29}$$

The dot product, found in the dual formulation, between the instance vectors appears both in the optimization function 28 and the classification function 29. Such a dot product can be replaced by the so called kernel functions [26]. Various kernel representations exists, however in this study, for the sake of clarity and generality, we are going to prove the concept of the method using polynomial kernels, defined in Eq. 30, which are known to be successful off-the-shelf kernels [26].

$$K([U_{i,*}, B_{U_i}], [U_{t,*}, B_{U_l}]) = \left(B_{U_i} B_{U_l} + \sum_{k=1}^{d} U_{i,k} U_{l,k} + 1 \right)^p \tag{30}$$

The ultimate objective function that defines nonlinear supervised dimensionality reduction is presented in Eq. 31. *This model, in cooperation with the forthcoming learning algorithm, are the main contributions of our paper.*

$$\underset{U,V,\alpha,B_U,B_V}{\text{argmin}} \quad F(X,Y,U,V,\alpha) = \beta \sum_{i=1}^{n+n'} \sum_{j=1}^{m} \left(X_{i,j} - \left(\sum_{k=1}^{d} U_{i,k} V_{k,j} + B_{U_i} + B_{V_j} \right) \right)^2$$

$$+ (1-\beta) \left(\frac{1}{2} \sum_{i=1}^{n} \sum_{l=1}^{n} \alpha_i \alpha_l Y_i Y_l \, K([U_{i,*}, B_{U_i}], [U_{t,*}, B_{U_l}]) - \sum_{i=1}^{n} \alpha_i \right)$$

$$+ \lambda_U \sum_{i=1}^{n+n'} \sum_{k=1}^{d} U_{i,k}^2 + \lambda_V \sum_{k=1}^{d} \sum_{j=1}^{m} V_{k,j}^2 \tag{31}$$

$$\text{s.t:} \quad 0 \le \alpha_i \le C, \quad i = 1, \dots, n$$

$$\sum_{i=1}^{n} \alpha_i Y_i = 0$$

Meanwhile the classification of a test instance U_t using kernels and the learned U, α, resulting from the solution of the dual joint optimization is shown in Eq. 32.

$$Y_t = \text{sign} \left(\sum_{i=1}^{n} \alpha_i Y_i \, K([U_{i,*}, B_{U_i}], [U_{t,*}, B_{U_l}]) + W_0 \right) \tag{32}$$

The Benefit of Non-linear Supervision is the ability to preserve both the reconstruction and the classification accuracy. This dual objective is achieved best if there is no sacrifice in terms of reconstruction. More concretely, let us assume the original data is non-linearly separable. Then, a linearly supervised decomposition cannot easily minimize both F_R and F_{CA}. The handicap is created due to trying to classify the low-rank data linearly, even though the original data is non-linear. As a consequence, the structure of the data cannot be accurately preserved and the reconstruction is poor, i.e. high F_R error. Unable to preserve the structure of the data, a linearly supervised decomposition struggles to achieve a competitive generalization of prediction accuracy over the test instances.

Figure 2 illustrates the benefit of the non-linear supervision with a concrete experiment. A 2-dimensional synthetic non-linearly separable dataset is created in sub-figure a). For experimental purposes we added noise through a new variable X_3 that contains random values between $[-1, 1]$. The key aspect of the experiment is to project the noisy 3-dimensional data back to 2-dimensions using both linearly (c)) and non-linearly (d)) supervised reductions. The projection parameters are found in the caption comment. As can be seen from sub-figure (c), the linear supervision cannot linearly separate all instances in the low-rank space under reasonable β values. On the contrary, a non-linear decomposition can achieve a 0% training error, because a non-linear arrangement of the data is easily achieved in the low-rank space. Please note that reasonable switch parameter values are $\beta > 0$. In the absurd case of $\beta = 0$ no reconstruction loss updates will be applied and the classification loss term will create a low-rank arrangement of the training instances without preserving at all the structure of the original data. Such a classifier is destined to under-perform over the test data.

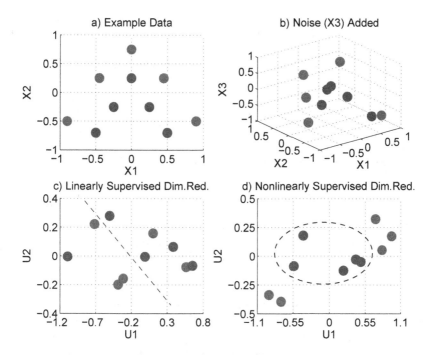

Fig. 2. Nonlinearly supervised reduction: (a) original data with two classes (blue and red); (b) random noise variable (X3) added; (c) linearly supervised dimensionality reduction (from (b) to (c)); (d) non-linearly supervised dimensionality reduction (from (b) to (d)). Parameters: $\eta_R = \eta_{CA} = 0.001$, $\lambda_U = \lambda_V = 0.01$, $\lambda_W = 2$, $C = 0.5$, $p = 3$ and $\beta = 0.7$ and 300 iterations (Color figure online).

3.6 Algorithm for Learning the Non-linearly Supervised Dimensionality Reduction

The objective function of Eq. 31 is a non-convex function in terms of U, V and W, which makes it challenging for optimization. However stochastic gradient descent is shown to perform efficiently in minimizing such non-convex functions [15]. The benefits of stochastic gradient descent rely on better convergence, because cells of X are randomly picked for optimization, thus updating different rows of U, instead of iterating through the all the features of the same instance.

On the other side, the classification accuracy terms of Eq. 28 can be solved, in terms of α, by any standard SVMs dual solver method in case we consider U to be fixed. Thus, in an alternating fashion we solve the α-s by keeping U fixed. Then in the next step we update U using the learned α-s and V matrix, by taking a step in the negative direction of the overall loss w.r.t U. The update of V is performed as last step. Those three steps can be repeated until convergence as shown in the Algorithm 3.

Similarly, we can split up the classification accuracy loss term, F_{CA}, into smaller loss terms $F_{CAi,l}$, defined per each instance pair (i, l).

$$F_{CA}(Y, U, \alpha) = \sum_{i=1}^{n} \sum_{l=1}^{n} F_{CA}(Y, U, \alpha)_{i,l} \tag{33}$$

$$F_{CA}(Y, U, \alpha)_{i,l} = (1 - \beta) \left(\frac{\alpha_i \alpha_l}{2} Y_i Y_l \, K([U_{i,*}, B_{U_i}], [U_{l,*}, B_{U_l}]) - \frac{\alpha_i + \alpha_l}{n^2} \right) \tag{34}$$

Gradients:

$$\frac{\partial F_{CA}(Y, U, \alpha)_{i,l}}{\partial U_{i,k}} = \frac{\beta - 1}{2} \alpha_i \alpha_l Y_i Y_l \, p \left(B_{U_i} B_{U_l} + \sum_{k=1}^{d} U_{i,k} U_{l,k} + 1 \right)^{p-1} U_{l,k} \tag{35}$$

$$\frac{\partial F_{CA}(Y, U, \alpha)_{i,l}}{\partial U_{l,k}} = \frac{\beta - 1}{2} \alpha_i \alpha_l Y_i Y_l \, p \left(B_{U_i} B_{U_l} + \sum_{k=1}^{d} U_{i,k} U_{l,k} + 1 \right)^{p-1} U_{i,k} \tag{36}$$

$$\frac{\partial F_{CA}(Y, U, \alpha)_{i,l}}{\partial B_{U_i}} = \frac{\beta - 1}{2} \alpha_i \alpha_l Y_i Y_l \, p \left(B_{U_i} B_{U_l} + \sum_{k=1}^{d} U_{i,k} U_{l,k} + 1 \right)^{p-1} B_{U_l} \tag{37}$$

$$\frac{\partial F_{CA}(Y, U, \alpha)_{i,l}}{\partial B_{U_l}} = \frac{\beta - 1}{2} \alpha_i \alpha_l Y_i Y_l \, p \left(B_{U_i} B_{U_l} + \sum_{k=1}^{d} U_{i,k} U_{l,k} + 1 \right)^{p-1} B_{U_i} \tag{38}$$

The updates of α-s is carried through an algorithm which is a reduced version of the Sequential Minimal Optimization (SMO) [27]. Since the dual form optimization function contains the constraint $\sum_{i=1}^{n} \alpha_i Y_i = 0$, then any update of an α_i will violate the constraint. Therefore SMO updates the α-s in pair, offering three heuristics which defines which subset of the pairs should be updates first, in order to speed up the algorithm.

In difference to the original algorithm, we have ignored the selection heuristic for the α pairs to update. The reason for omitting the heuristics is due to the fact that U instances are continuously updated/modified. For instance, let us consider an imaginary instance U_i far away from the decision boundary, which means $\alpha_i = 0$. However in the next iteration, the instance U_i might be updated and move close to the boundary, meaning that α_i becomes a candidate for being updated ($0 < \alpha_i \leq C$), opposite to the functioning of SMO heuristic that would have avoided updating the instance, alluding that α_i is still 0.

The alpha updates rely on solving the function analytically for a pair of α-s at a step, until no $\alpha_i, \forall i$, violates the KKT [27] conditions described in Eq. 39.

$$\text{Let } \hat{Y}_i = \text{sign} \left(\sum_{j=1}^{n} \alpha_j Y_j \, K([U_{i,*}, B_{U_i}], [U_{t,*}, B_{U_l}]) + W_0 \right)$$

$$\alpha_i = 0 \rightarrow Y_i \hat{Y}_i \geq 1$$
$$0 < \alpha_i < C \rightarrow Y_i \hat{Y}_i = 1$$
$$\alpha_i = C \rightarrow Y_i \hat{Y}_i \leq 1 \tag{39}$$

Therefore the learning algorithm will update all the pairs of α-s in each iteration. The SMO-like update of each pair of alphas is shown in the Algorithm 2, with more details in [27]. Please note that the algorithm also updates the hyperplane intercept W_0, which is used for classification of latent instances.

Algorithm 2. UpdateAlphaPair

Input: First alpha index i, Second alpha index j
Output: Updated α and W_0

$(\alpha_i^{old}, \alpha_j^{old}) \leftarrow (\alpha_i, \alpha_j)$

Let $s \leftarrow Y_i Y_j$

$(L, H) \leftarrow \left(\max(0, \alpha_j^{old} + s\alpha_i^{old} - \frac{s+1}{2}C), \min(C, \alpha_j^{old} + s\alpha_i^{old} - \frac{s-1}{2}C) \right)$

$E_k \leftarrow \left(\sum_{l=0}^{n} Y_l \alpha_l K(U_{l,*}, U_{k,*}) + W_0 \right) - Y_k, \forall k \in \{i, j\}$

$\alpha_j^{new} \leftarrow \alpha_j^{old} - \frac{Y_j(E_i - E_j)}{2K(U_{i,*}, U_{j,*}) - K(U_{i,*}, U_{i,*}) - K(U_{j,*}, U_{j,*})}$

$\alpha_j^{new,clipped} = \begin{cases} L, & \text{if } \alpha_j^{new} < L \\ \alpha_j^{new}, & \text{if } L < \alpha_j^{new} < H \\ H, & \text{if } \alpha_j^{new} > H \end{cases}$

$\alpha_i^{new} \leftarrow \alpha_i^{old} + s(\alpha_j^{new,clipped} - \alpha_j^{old})$

$b_i \leftarrow E_i + y_i(\alpha_i^{new} - \alpha_i^{old})K(U_{i,*}, U_{i,*}) + Y_2(\alpha_j^{new,clipped} - \alpha_j^{old})K(U_{i,*}, U_{j,*}) + W_0$

$b_j \leftarrow E_j + y_i(\alpha_i^{new} - \alpha_i^{old})K(U_{i,*}, U_{i,*}) + Y_2(\alpha_j^{new,clipped} - \alpha_j^{old})K(U_{i,*}, U_{j,*}) + W_0$

$\mathbf{W_0} \leftarrow \frac{b_i + b_j}{2}, (\alpha_j, \alpha_i) \leftarrow \left(\alpha_j^{new,clipped}, \alpha_i^{new} \right)$

return α, W_0

Having defined the gradients for updating latent matrices U, V with respect to the optimization loss and also the update rules for α-s, we can derive a final learning algorithm based on coordinate gradient descent. Algorithm 3 shows the learning algorithm in full terms. The updates of each cell of U, V, B_U, B_V, as response to the reconstruction loss F_R and the classification accuracy loss F_{CA}, are conducted in the negative direction of the gradients scaled by hyper-parameter learning rates η_R, η_{CA}. The convergence is guaranteed by selecting small values for the learning rates. The stopping criteria is when the final loss from Eq. 31 reaches an optimum, meaning it doesn't get further minimized.

The Convergence of Learning is guaranteed because both steps (i) the learning of the latent data U, V for the reconstruction loss, (ii) updates of U for the classification loss and the Lagrangian multipliers, are both steps of the Expectation Maximization algorithm. Figure 3 specifically illustrate the convergence of our algorithm for the Ionosphere dataset.

The reconstruction loss (F_R) and the classification accuracy loss (F_{CA}) converge smoothly as depicted in the left plot of Fig. 3. The learning algorithm updates the latent data with respect to the objective function of Eq. 1, therefore the decrease of the values of loss terms (shown in sub-figure (a)) is an indication that our algorithm converge as expected. On the right, sub-figure (b) demonstrates the consequence that a minimization of the classification loss has towards decreasing the error rate on both training and testing data. There is no significant gap between the train and test errors, which indicates that the hyper-plane (α) learned over training instances generalizes accurately on the unobserved test data.

Algorithm 3. Learning Algorithm: Nonlinearly Supervised Dim. Red.

Input: Dataset matrix $X \in \mathbb{R}^{(n+n') \times m}$, Labels vector $Y \in \mathbb{R}^n$, **Parameters:** { Box
    constraint C, Optimization switch β, Latent dimensions d, Learning rates η_R, η_{CA},
    Regularizations λ_U, λ_V, Kernel degree p }
Output: $U, V, B_U, B_V, \alpha, W_0$
    Initialize $U \in \mathbb{R}^{(n+n') \times d}$, $V \in \mathbb{R}^{d \times m}$, $B_U \in \mathbb{R}^{(n+n') \times 1}$, $B_V \in \mathbb{R}^{1 \times m}$ randomly
    Initialize $\alpha \leftarrow \{0\}^n$, $W_0 \leftarrow 0$
    while F **not** reached an optimum **do**
        for $\forall (i, j, k) \in (\{1...(n+n')\}, \{1...m\}, \{1...d\})$ *in random order* **do**
            $U_{i,k} \leftarrow U_{i,k} - \eta_R \frac{\partial F_R(X,U,V)_{i,j}}{\partial U_{i,k}}$
            $V_{k,j} \leftarrow V_{k,j} - \eta_R \frac{\partial F_R(X,U,V)_{i,j}}{\partial V_{k,j}}$
            $B_{U_i} \leftarrow B_{U_i} - \eta_R \frac{\partial F_R(X,U,V)_{i,j}}{\partial B_{U_i}}$
            $B_{V_j} \leftarrow B_{V_j} - \eta_R \frac{\partial F_R(X,U,V)_{i,j}}{\partial B_{V_j}}$
        end for
        for $\forall (i, l, k) \in (\{1...n\}, \{1...n\}, \{1...d\})$ *in random order* **do**
            $U_{i,k} \leftarrow U_{i,k} - \eta_{CA} \frac{\partial F_{CA}(Y,U,\alpha)_{i,l}}{\partial U_{i,k}}$
            $U_{l,k} \leftarrow U_{l,k} - \eta_{CA} \frac{\partial F_{CA}(Y,U,\alpha)_{i,l}}{\partial U_{l,k}}$
            $B_{U_i} \leftarrow B_{U_i} - \eta_{CA} \frac{\partial F_{CA}(Y,U,\alpha)_{i,l}}{\partial B_{U_i}}$
            $B_{U_l} \leftarrow B_{U_l} - \eta_{CA} \frac{\partial F_{CA}(Y,U,\alpha)_{i,l}}{\partial B_{U_l}}$
        end for
        for $\forall i \in \{1 \ldots n\}$ **do**
            if α_i violates KKT of Equation 40 **then**
                for $\forall j \in \{1 \ldots n\}$ *in random order* **do**
                    $(\alpha, W_0) \leftarrow$ UpdateAlphaPair(i, j), from Algorithm 2
                end for
            end if
        end for
    end while
    return $U, V, B_U, B_V, \alpha, W_0$

The Algorithmic Complexity of our method both in terms of run-time and space depends on the size of the data. Concretely, the storage requirements are upper bounded to the size of the predictors, since for $d < m$ the storage of Y, U, V and α are all less than X. Therefore the space complexity of the method is $O((n + n') \times m)$. The running time depends on the number of iterations of Algorithm 3. If we denote the iterations as I, then the number of updates is proportional to $O(I \times (n + n') \times m \times d)$, since for every cell $X_{i,j}$ we update all the k-many row cells $U_{i,k}$ and column cells $V_{j,k}$. The updates of the classification accuracy loss term α-s are inferior in number ($O(I \times (n + n') \times m)$) and do not influence the upper bounding algorithmic complexity with respect to the reconstruction loss. Note that l, the number of target categories, is a small constant equals to two for binary problems. It is not possible to forecast exactly the number of iterations that a dataset will require until convergence, since it

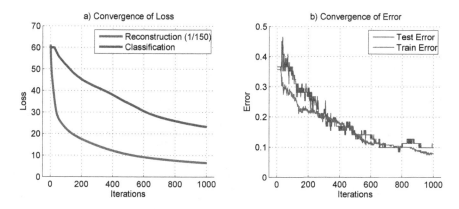

Fig. 3. Convergence of the loss terms and errors belonging to the Ionosphere dataset; Parameters $\beta = 0.7$, $C = 1, p = 1$, $\lambda_U = \lambda_V = 10^{-5}$, $\eta_R = 0.001, \eta_{CA} = 0.0001$ and $d = 25$.

depends on the slope of the loss function's surface $(F_R + F_{CA})$ and the learning rate parameter. Large learning rates converge faster but increase both the risk of divergence and missing narrow local optima. On the other hand, small learning rates require more iterations to minimize the loss.

4   Experimental Results

4.1   Experimental Setup

In order to compare the classification accuracy of our method Nonlinearly Supervised Dimensionality Reduction (**NSDR**), we implemented and compared against three baselines:

- **PCA-SVMs:** Matcshing against the standard PCA dimensionality reduction and then SVMs classification will demonstrate the advantage of supervised decomposition against unsupervised decomposition (PCA).
- **SVMs:** Comparison against the default SVMs will provide insights on the advantages of dimensionality reduction.
- **LSDR:** The linearly supervised dimensionality reduction represents the state of the art in data projection. A comparison against this baseline clarifies the superiority position of our method [3,4]. The smooth hinge loss is used due to its competitiveness and high prediction accuracy demonstrated together with SVMs.

The experiments were conducted using five folds cross validation, where the data was divided into five splits and each split was, in turn, the test and the other four the training data.

The hyper parameters of our method and the baselines was selected using a validation data split from the training data. The best grid-search combinations of hyper parameters that yielded the best accuracy was selected for

Table 1. Hyper-parameter Search Results

DATASET	LSDR	NSDR	PCA-SVMs	SVMs
breast_canc.	$\lambda_U = 10^{-4}; \lambda_V = 10^{-5};$ $\eta_R = 10^{-3}; \eta_{CA} = 10^{-4}; d = 6;$ $\beta = 0.1, \lambda_W = 1$	$\lambda_U = 10^{-2}; \lambda_V = 1;$ $\eta_R = 10^{-3}; \eta_{CA} = 10^{-4}; d = 9;$ $\beta = 0.9; C = 10; p = 2$	$var = 1;$ $C = 10; p = 2$	$C = 10$ $p = 2$
ionosphere	$\lambda_U = 10^{-6}; \lambda_V = 10^{-5};$ $\eta_R = 10^{-3}; \eta_{CA} = 10^{-4}; d = 25;$ $\beta = 0.9, \lambda_W = 1$	$\lambda_U = 10^{-6}; \lambda_V = 10^{-6};$ $\eta_R = 10^{-3}; \eta_{CA} = 10^{-4}; d = 9;$ $\beta = 0.9; C = 10; p = 2$	$var = 1;$ $C = 1; p = 3$	$C = 0.1$ $p = 2$
pi-diabetes	$\lambda_U = 10^{-6}; \lambda_V = 10^{-3};$ $\eta_R = 10^{-3}; \eta_{CA} = 10^{-4}; d = 6;$ $\beta = 0.9, \lambda_W = 0.1$	$\lambda_U = 10^{-4}; \lambda_V = 1;$ $\eta_R = 10^{-3}; \eta_{CA} = 10^{-4}; d = 6;$ $\beta = 0.1; C = 0.1; p = 3$	$var = 1$ $C = 1; p = 3$	$C = 10$ $p = 3$
sonar	$\lambda_U = 10^{-6}; \lambda_V = 1;$ $\eta_R = 10^{-3}; \eta_{CA} = 10^{-4}; d = 45;$ $\beta = 0.5, \lambda_W = 10$	$\lambda_U = 10^{-2}; \lambda_V = 1;$ $\eta_R = 10^{-3}; \eta_{CA} = 10^{-4}; d = 60;$ $\beta = 0.1; C = 0.1; p = 2$	$var = 0.7$ $C = 10; p = 2$	$C = 0.1$ $p = 3$
spect	$\lambda_U = 10^{-2}; \lambda_V = 1;$ $\eta_R = 10^{-3}; \eta_{CA} = 10^{-4}; d = 16;$ $\beta = 0.5, \lambda_W = 0.1$	$\lambda_U = 10^{-2}; \lambda_V = 10^{-5};$ $\eta_R = 10^{-3}; \eta_{CA} = 10^{-4}; d = 22;$ $\beta = 0.5; C = 0.1; p = 3$	$var = 1$ $C = 1; p = 3$	$C = 0.1$ $p = 2$

being applied to the test split. The ranges of search for the LSDR and NSDR methods were $\lambda_U \in \{10^{-6}, 10^{-5}, \dots, 10^0, 10^1\}$, $\lambda_V \in \{10^{-6}, 10^{-5}, \dots, 10^0, 10^1\}$, $\eta_R \in \{10^{-4}, 10^{-3}\}, \eta_{CA} \in \{10^{-4}, 10^{-3}\}$, $d \in \{25\%, 50\%, 75\%, 100\%\}$ of m, $\beta \in \{0.1, 0.5, 0.9\}$, $C \in \{0.1, 1, 10\}, p \in \{1, 2, 3, 4\}$. For PCA-SVMs there is a variance parameter $var \in \{0.5, 0.7, 1.0\} \times 100\%$. The other SVMs parameters C, p for both PCA-SVMs and SVMs were searched in the same ranges as the ones reported for NSDR previously.

4.2   Results

In order to validate the proposed method we selected five popular binary datasets from the UCI repository, which cover a range of applications such as medicine (Breast Cancer, Pi-Diabetes and Spect), radar (Ionosphere) and undersea explorations (Sonar). The hyper-parameters of the models were fit in a **5-folds** cross-validation fashion among the aforementioned ranges. In order to promote experimental reproducibility, we present the exact values of parameters in Table 1 for our method and all the baselines.

The Sensitivity of Parameters depends on the impact that a perturbation of the value of a parameter has over the error rate. The most sensitive parameters in a learning method are usually the regularization weights of the model complexity. In the case of NSDR, the learning rate, the number of iterations and the latent dimensions are less critical with respect to accuracy. The learning rate should be set small enough to avoid divergence and the iterations large enough to ensure convergence. In order to avoid under-fitting, the number of dimensions has to be set at a large value, e.g. 75 % of m. Therefore, the most sensitive parameters are λ_U, λ_V for the reconstruction loss and C, p for the classification accuracy loss. Figure 4 illustrate the sensitivity relation among the complexity regularization parameters on the SPECT dataset. The left plot demonstrate the error rate heatmap (the smaller the less error) as a result of perturbing the values of λ_U

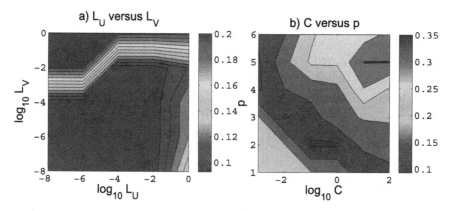

Fig. 4. Parameter Sensitivity Analysis on SPECT dataset; Parameters $\beta = 0.9$, $C = 0.1$, $p = 2$, $\lambda_U = 10^{-6}$, $\lambda_V = 10^{-5}$, $\eta_R = 0.001$, $\eta_{CA} = 0.0001$ and $d = 11$.

versus λ_V. All the other parameters are kept constant at their optimal value (yielding smallest error) and are displayed in the figure caption. The ranges of the plot axis are displayed as the logarithm of the parameters values in order to have equidistant ticks. The plot on the right presents the sensitivity of the error rate with respect to the changes of the accuracy parameters C and p. As can be seen from both plots, the error fluctuates significantly in the case of C, p, which indicates that a practitioner should search for those parameters in a narrow grid of values. On the contrary, the search for λ_U, λ_V are less critical because there exists a large region of optimal values, which is expressed as a blue plateau.

We would like to point out that our non-linearly supervised dimensionality reduction (NSDR) is a generalization of the linearly supervised projection (LSDR). The linear case can be instantiated as a polynomial kernel of degree one. Since our method NSDR includes the functionality of LSDR, then every competitive result of LSDR is easily achieved by NSDR (with parameter $p = 1$). On the contrary, as our experiments show, a linear decomposition does not recover the inexpressive nature of its linear hyper-plane.

The accuracy results in terms of error ratios is presented in Table 2 with respect to five real-life datasets. The winning method per each dataset is shown in **bold**. As we can observe our proposed method outperforms the baselines in all the datasets.

NSDR improves the classification on the *ionosphere* and *sonar* datasets with significant differences, while on the other datasets the gap to the second best is smaller. As can be deduced from the results, the linearly supervised decomposition is superior to the unsupervised decomposition (PCA-SVMs) in 3 out of 5 datasets. Furthermore the nonlinear supervision (NSDR) outperforms the linear method (LSDR) in all the datasets. The outcomes of the experiments validate the expectations of our paper and demonstrate the usefulness of non-linear supervision with respect to real-life data.

Table 2. Error Ratios on Real-Life Datasets

DATASET	NSDR	LSDR	PCA-SVMs	SVMs
breast_cancer_w	**0.070 ± 0.018**	0.122 ± 0.013	0.082 ± 0.019	0.073 ± 0.021
ionosphere	**0.066 ± 0.008**	0.097 ± 0.016	0.091 ± 0.010	0.140 ± 0.018
pi-diabetes	**0.264 ± 0.023**	0.279 ± 0.016	0.280 ± 0.006	0.274 ± 0.030
sonar	**0.106 ± 0.041**	0.188 ± 0.042	0.226 ± 0.129	0.226 ± 0.056
spect	**0.138 ± 0.051**	0.142 ± 0.056	0.243 ± 0.103	0.206 ± 0.002

4.3   Run-Time Disadvantage

While our proposed method (NSDR) achieves a better classification accuracy than the baseline, still it has a costly optimization procedure. Compared to faster classifiers like SVMs, our method has a joint factorization and classification loss term. The joint optimization requires significant time to compute, in particular because of the slower learning rates that are required to ensure a convergence. For instance, it takes only 0.415 s for the SVMs and 158.065 for NSDR to compute. Clearly, the run-time is a negative aspect of the paper with respect to SVMs. As a rule of thumb, we advice practitioners to use NSDR only if classification accuracy, not run-time, is the primary objective.

5   Conclusions

Throughout this study we presented a non-linearly supervised dimensionality reduction technique, which jointly combined a joint optimization on reconstruction and classification accuracy. Such an approach distances from traditional data mining that considered dimensionality reduction and classification as two disjoint, sequential processes. The supervised decomposition benefits from the knowledge on the target values of training instances, in order to both eliminate the noise present in the predictor values and also preserve the class segregation.

In our presented method, the reconstruction loss term is expressed as matrix factorization decomposition of latent matrices, while the classification accuracy as a dual form kernel maximum margin classifier. Consequently, the reduced dataset is learned via a coordinate descent algorithm which updates the reduced dimensionality dataset w.r.t to both loss terms simultaneously.

Existing state of the art methods in supervised decomposition incorporate linear classification terms in the objective function. In contrast, our method introduces a novel non-linear supervision of the dimensionality reduction process. We adopt a kernel based classification loss, which guides the low-rank data into being separated via a non-linear hyper-plane. A non-linear decomposition improves accuracy in cases where the original data is not linearly separable, because preserving the non-linear arrangement of instances does not deteriorate the reconstruction loss of predictor values. Since the linear supervision is a special instance of our methods for a polynomial kernel of degree, then our method offers a super-set of expressiveness.

Empirical results over five real-life datasets show that the proposed method outperforms the selected baselines in the majority of the datasets. Significant improvement is present against unsupervised techniques, which indicates the benefit of incorporating target value information into dimensionality reduction. In addition, experimental results validated the superiority of non-linearly guided supervision against the linearly supervised state of the art decomposition.

Acknowledgment. This study was funded by the Seventh Framework Programme (FP7) of the European Commission, through projects REDUCTION(www.reduction-project.eu) and iTalk2Learn(www.italk2learn.eu).

In addition, the authors express their gratitude to Lucas Rego Drumond (University of Hildesheim) for his assistance on formalizing the linearly supervised decomposition.

References

1. Samet, H.: Foundations of Multidimensional and Metric Data Structures (The Morgan Kaufmann Series in Computer Graphics and Geometric Modeling). Morgan Kaufmann Publishers Inc., San Francisco (2005)
2. Grabocka, J., Bedalli, E., Schmidt-Thieme, L.: Efficient classification of long time-series. In: Markovski, S., Gushev, M. (eds.) ICT Innovations 2012. AISC, vol. 207, pp. 47–57. Springer, Heidelberg (2013)
3. Grabocka, J., Nanopoulos, A., Schmidt-Thieme, L.: Classification of sparse time series via supervised matrix factorization. In: Hoffmann, J., Selman, B. (eds.) AAAI, AAAI Press (2012)
4. Das Gupta, M., Xiao, J.: Non-negative matrix factorization as a feature selection tool for maximum margin classifiers. In: Proceedings of the 2011 IEEE Conference on Computer Vision and Pattern Recognition, CVPR 2011, pp. 2841–2848. IEEE Computer Society, Washington, DC (2011)
5. Jolliffe, I.T.: Principal Component Analysis, 2nd edn. Springer, New York (2002)
6. Wismüller, A., Verleysen, M., Aupetit, M., Lee, J.A.: Recent advances in nonlinear dimensionality reduction, manifold and topological learning. In: ESANN (2010)
7. Hoffmann, H.: Kernel pca for novelty detection. Pattern Recognit. **40**(3), 863–874 (2007)
8. Sun, J., Crowe, M., Fyfe, C.: Extending metric multidimensional scaling with bregman divergences. Pattern Recognit. **44**(5), 1137–1154 (2011)
9. Gorban, A.N., Zinovyev, A.Y.: Principal manifolds and graphs in practice: from molecular biology to dynamical systems. Int. J. Neural Syst. **20**(3), 219–232 (2010)
10. Lee, J.A., Verleysen, M.: Nonlinear Dimensionality Reduction. Springer, New York; London (2007)
11. Gashler, M.S., Martinez, T.: Temporal nonlinear dimensionality reduction. In: Proceedings of the IEEE International Joint Conference on Neural Networks, IJCNN 2011, pp. 1959–1966. IEEE Press (2011)
12. Lawrence, N., Hyvrinen, A.: Probabilistic non-linear principal component analysis with gaussian process latent variable models. J. Mach. Learn. Res. **6**, 1783–1816 (2005)
13. Lawrence, N.: Gaussian process latent variable models for visualisation of high dimensional data. In: NIPS (2003, 2004)

14. Singh, A.P., Gordon, G.J.: A unified view of matrix factorization models. In: Daelemans, W., Goethals, B., Morik, K. (eds.) ECML PKDD 2008, Part II. LNCS (LNAI), vol. 5212, pp. 358–373. Springer, Heidelberg (2008)

15. Koren, Y., Bell, R.M., Volinsky, C.: Matrix factorization techniques for recommender systems. IEEE Comput. **42**(8), 30–37 (2009)

16. Rendle, S., Schmidt-Thieme, L.: Online-updating regularized kernel matrix factorization models for large-scale recommender systems. In: Pu, P., Bridge, D.G., Mobasher, B., Ricci, F. (eds.) RecSys, pp. 251–258. ACM (2008)

17. Cai, D., He, X., Han, J., Huang, T.S.: Graph regularized nonnegative matrix factorization for data representation. IEEE Trans. Pattern Anal. Mach. Intell. **33**(8), 1548–1560 (2011)

18. Giannakopoulos, T., Petridis, S.: Fisher linear semi-discriminant analysis for speaker diarization. IEEE Trans. Audio Speech Lang. Process. **20**(7), 1913–1922 (2012)

19. Menon, A.K., Elkan, C.: Predicting labels for dyadic data. Data Min. Knowl. Discov. **21**(2), 327–343 (2010)

20. Rish, I., Grabarnik, G., Cecchi, G., Pereira, F., Gordon, G.J.: Closed-form supervised dimensionality reduction with generalized linear models. In: ICML 2008: Proceedings of the 25th International Conference on Machine Learning, pp. 832–839. ACM, New York (2008)

21. Rennie, J.D.M.: Loss functions for preference levels: regression with discrete ordered labels. In: Proceedings of the IJCAI Multidisciplinary Workshop on Advances in Preference Handling, pp. 180–186 (2005)

22. Fukumizu, K., Bach, F.R., Jordan, M.I.: Dimensionality reduction for supervised learning with reproducing kernel hilbert spaces. J. Mach. Learn. Res. **5**, 73–99 (2004)

23. Salakhutdinov, R., Hinton, G.: Learning a nonlinear embedding by preserving class neighbourhood structure. In: Proceedings of the International Conference on Artificial Intelligence and Statistics, vol. 11 (2007)

24. Zhang, D., Zhou, Z.-H., Chen, S.: Semi-supervised dimensionality reduction. In: Proceedings of the 7th SIAM International Conference on Data Mining, pp. 11–393 (2007)

25. Urtasun, R., Darrell, T.: Discriminative gaussian process latent variable models for classification. In: International Conference in Machine Learning (2007)

26. Scholkopf, B., Smola, A.J.: Learning with Kernels: Support Vector Machines, Regularization, Optimization, and Beyond. MIT Press, Cambridge (2001)

27. Platt, J.: Fast training of support vector machines using sequential minimal optimization. In: Schoelkopf, B., Burges, C., Smola, A. (eds.) Advances in Kernel Methods - Support Vector Learning. MIT Press, Cambridge (1998)

Metrics for Association Rule Clustering Assessment

Veronica Oliveira de Carvalho[1]([⊠]), Fabiano Fernandes dos Santos[2],
and Solange Oliveira Rezende[2]

[1] Instituto de Geociências e Ciências Exatas,
UNESP - Universidade Estadual Paulista, Rio Claro, Brazil
`veronica@rc.unesp.br`
[2] Instituto de Ciências Matemáticas e de Computação,
USP - Universidade de São Paulo, São Carlos, Brazil
`{fabianof,solange}@icmc.usp.br`

Abstract. Issues related to association mining have received attention, especially the ones aiming to discover and facilitate the search for interesting patterns. A promising approach, in this context, is the application of clustering in the pre-processing step. In this paper, eleven metrics are proposed to provide an assessment procedure in order to support the evaluation of this kind of approach. To propose the metrics, a subjective evaluation was done. The metrics are important since they provide criteria to: (a) analyze the methodologies, (b) identify their positive and negative aspects, (c) carry out comparisons among them and, therefore, (d) help the users to select the most suitable solution for their problems. Besides, the metrics do the users think about aspects related to the problems and provide a flexible way to solve them. Some experiments were done in order to present how the metrics can be used and their usefulness.

Keywords: Association rules · Pre-processing · Clustering · Evaluation metrics

1 Introduction

Association has been highlighted, among the data mining tasks, due to its comprehensibility even by non-experts in the field. To have an idea, the *Apriori* algorithm, broadly used to obtain association patterns, was elected as one of the ten data mining algorithms most employed by the community [1]. For this and other reasons, association has been applied in many domains, as noticed in [2–6] works.

In the last years, researches have adopted some strategies to aid the user to identify the relevant associative patterns of the domain. One of these strategies is to pre-process the data before obtaining the rules. For that, many approaches have been proposed, being clustering a promising one. In this case, as seen in Fig. 1, the data are initially grouped into n groups ($GD_1,GD_2,...,GD_n$). Association rules are extracted within each group and, in the end, n groups of rules are

© Springer-Verlag Berlin Heidelberg 2015
A. Hameurlain et al. (Eds.): TLDKS XVII, LNCS 8970, pp. 97–127, 2015.
DOI: 10.1007/978-3-662-46335-2_5

obtained $(GR_1, GR_2, ..., GR_n)$. All these rules compose the rule set. According to [7], each group expresses its own associations without the interference of the other groups that contain different association patterns. The aim is to obtain potentially interesting rules that would not be extracted from unpartitioned data sets[1]. The user must set the minimum support to a low value to discover these same patterns from unpartitioned data sets, causing a rapidly increase in the number of rules. Recent works have used these ideas in different domains as [8] in maintenance systems, [9] in banking context and [10,11] in automotive data.

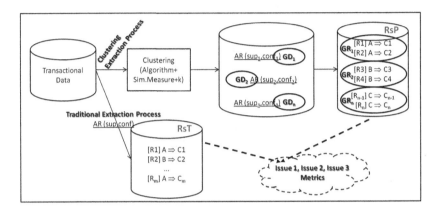

Fig. 1. Overview of the process to extract association rules through clustering in comparison to the traditional one.

Distinct methodologies have been proposed to enable the described process. Each methodology uses a different combination of similarity measures with clustering algorithms to obtain the groups of rules (see Fig. 1). However, little has been done to analyze the performance of the methodologies or even to compare the results. For example, given a specific problem of a domain, how could the user identify the most suitable methodology to use in his problem? Or, how the user could check if the selected methodology was good enough for the problem, considering that different interests may be important for his decision? To aid the user in these tasks, considering the aims of the clustering process above discussed, there are some issues that have to be investigated (see Fig. 1):

Issue 1. Is there overlap between a rule set obtained through partitioned data, i.e., extracted from clustered data, in relation to a rule set obtained through unpartitioned data, i.e., extracted from traditional process? A *rule set* obtained through a *partitioned* data is named here as RsP and a *rule set* obtained through a *traditional* process is named here as RsT.

---

[1] In this work, it is assumed that a pattern is interesting if it is relevant and/or useful to the user – rules having high support and/or high confidence are not necessarily interesting to the user.

Issue 2. Is there overlap between the rules in RsT and RsP regarding the interesting knowledge? In other words, has RsP, in fact, more interesting patterns than RsT?

Issue 3. What is the process behavior regarding the number of rules that are obtained in RsP?

Based on the exposed arguments and on the three presented issues, eleven metrics are proposed in this paper to provide an assessment procedure in order to support the evaluation of the methodologies that use clustering in the pre-processing step. However, to propose the metrics, a subjective evaluation was done with some users to understand each issue. The analysis was done as a way to comprehend the important aspects to be considered by a user when comparing the performance of some methodologies or even their results. Thereby, this paper contributes with current researches since the metrics provide criteria to: (a) analyze the methodologies, (b) identify their positive and negative aspects, (c) carry out comparisons among them and, therefore, (d) help the users to select the most suitable solution for their problems. Besides, the paper does the users think about many aspects to be considered in the presented context and provides to them a flexible way to explore their problems. This paper is an extension of [12]'s work, in which two more metrics are proposed (M_P, M_{RF}) and where the subjective evaluation is used to attest the importance of the previous obtained results.

This paper is organized as follows. Section 2 surveys the related researches. Section 3 presents the subjective evaluation and Sect. 4 the proposed metrics. Section 5 describes some experiments that were carried out to show how the metrics can be used. Section 6 discusses the results obtained in the experiments. Conclusion is given in Sect. 7.

2   Related Works

There are many researches that initially cluster data aiming to discover and facilitate the search for the interesting pattern of the domain. Some of these works are described below.

Plasse et al. [11] propose to split the transactions' items into groups in order to extract the rules. The aim is to find interesting associations among rare items (less frequent) that would not be discovered if the traditional process were applied, mainly in sparse data. Besides, the authors show that if the data are clustered it is possible to reduce both the amount and the complexity of obtained rules compared to non-clustered data. However, these results depend on the used clustering algorithms and similarity measures. The authors evaluate many hierarchical algorithms (Single, Complete, Average, Ward, Varclus) combined with many similarity measures (Jaccard, Russel Rao, Dice, Ochiai, Pearson). Nevertheless, it is not understandable how the rules are obtained within the groups, since it is necessary to have a set of transactions and not a set of items. This means that it is not clear how the transactions are distributed over the groups. Among the similarity measures used by them, we emphasize Jaccard (Eq. 1) and Russel Rao (Eq. 2) – the Russel Rao due to its good performance in the experiments presented in [11] and the Jaccard due to its use by the measure described below (Agg). The Jaccard between two items i_1 and i_2, named here as P-J (*Plasse Jaccard*), is the ratio between the transactions t the items cover simultaneously and the total of transactions the items cover. An item covers a transaction t if the item is in t. The Russel Rao between two items i_1 and i_2, named here as P-RR (*Plasse Russel Rao*), is computed considering the transactions t the items cover simultaneously over all the transactions. In fact, this measure is equivalent to the support of $i_1 \cap i_2$. In Eq. 2, N_t represents the total number of transactions.

$$\text{P-J}(i_1, i_2) = \frac{|\{t \text{ covered by } i_1\} \cap \{t \text{ covered by } i_2\}|}{|\{t \text{ covered by } i_1\} \cup \{t \text{ covered by } i_2\}|}. \tag{1}$$

$$\text{P-RR}(i_1, i_2) = \frac{|\{t \text{ covered by } i_1\} \cap \{t \text{ covered by } i_2\}|}{N_t}. \tag{2}$$

Aggarwal et al. [13] propose an algorithm, named CLASD, to split the transactions aiming to discover associations on small segments (subsets) of the data. The authors justify that the approach has a considerable impact on the rules that are obtained, since the patterns that cannot be identified in the whole set can be identified in the subsets. To cluster the transactions, [13] use a similarity measure proposed by them (Eq. 3), named here as Agg (*Aggarwal*). The clustering is done by CLASD. In this case, the similarity between two transactions t_1 and t_2 is computed by the affinity (Af) average among the transactions' items. It can be noted that the affinity (Af) is equivalent to the measure P-J. Therefore, after computing P-J among the m items in t_1 and the n items in t_2, the average among them is obtained. The higher the affinity among the items the more similar the transactions. From this initially clustering the rules are then extracted. Different from [11], the number of rules that are extracted by their process is higher compared to the traditional process, but smaller if the support in the traditional process had to be set with a very low value to obtain the same associations. A limitation of the approach is the number of parameters that has to be set (five plus two constants).

$$\text{Agg}(t_1, t_2) = \frac{\sum_{p=1}^{m} \sum_{q=1}^{n} Af(i_p, j_q)}{m * n}, \text{Af(i,j)} = \frac{sup(\{i, j\})}{sup(\{i\}) + sup(\{j\}) - sup(\{i, j\})}. \tag{3}$$

Koh and Pears [7] propose a methodology to cluster transactions and then extract a rare association rule set. The algorithm they proposed, named *Apriori Inverse*, applies upon groups of transactions in order to verify if it is possible to extract rare rules that are not discovered considering the whole data. Therefore, the idea is similar to the works above described. The authors demonstrate that the clustering contributes to the discovery of new associations. To cluster the transactions, their algorithm initially finds k seeds (centroids), where k indicates the number of frequent itemsets in transactions that match some conditions. Each seed forms a group. After the seed generation, each transaction t is allocated to one of the groups (seed) considering the similarity obtained through the measure presented in Eq. 4. The equation is an adaptation of the Jaccard coefficient and it compares the number of common items that occur between the transaction (t) and the group centroid (c_k). In this case, the higher the overlap between t and a centroid the higher the similarity value (Sim = similarity). In this approach it is necessary to set two additional values beyond the minimum support, hindering a little the exploration.

$$Sim(t, c_k) = \frac{|t \cap c_k|}{|t \cup c_k| - |t \cap c_k| + 1}. \tag{4}$$

There are other researches concerned with the clustering of transactions that, although not related to the extraction of association rules, could be used for that purpose [14–17]. In [18] the authors propose an approach to identify, a priori, the potentially interesting items to appear in the antecedents and in the consequents of the association rules without extracting them. The approach is divided in two steps: the clustering of the transactions and the selection of the interesting items. To do the clustering the authors propose the use of incremental K-means with the similarity measure presented in Eq. 5, named here as Denza. Note that this measure is the Jaccard between transactions. Therefore, the similarity between two transactions t_1 and t_2 is computed considering the items the transactions share. After the grouping, statistics are applied upon the groups to identify the items that are relevant to the application.

$$\text{Denza}(t_1, t_2) = \frac{|\{items\ in\ t_1\} \cap \{items\ in\ t_2\}|}{|\{items\ in\ t_1\} \cup \{items\ in\ t_2\}|}. \tag{5}$$

Among the papers above described, little has been done to analyze the performance of the methodologies, allowing to identify their positive and negative aspects, or even to compare the results among them. In general, the researchers compare the number of rules and/or itemsets that are obtained from unpartitioned data and clustered data to expose the usefulness of the methodologies. This strategy can be found in [7,11,13] and is related to "Issue 3" of Sect. 1. However, [11] also analyze the process considering the complexity of the rules that are obtained – the greater the number of items that compose a rule the higher its complexity. Koh and Pears [7] and Aggarwal et al. [13] discuss about some rules found through clustering to show that the process provides the discovery of interesting patterns, but the analysis of the process is subjective. Aggarwal et al. [13] also consider the execution time. Finally, [7] is the only work that allows a better analysis considering the existing overlap between the rules obtained from unpartitioned data and clustered data. This strategy is related to "Issue 1" of Sect. 1. Based on the presented arguments, as mentioned before, the necessity of an assessment procedure becomes evident.

It is important to mention that cluster validation indices (CVI), independently of being relative, internal, or external [19,20], are not reviewed here. It is understood that these CVI can be used to evaluate the groups of transactions before obtaining the association rules: in this case, the aim is to check if the partition fits the structure underlying the data. However, after obtaining the groups of transactions, the rules are obtained and, therefore, the CVI can no longer be used, since the data used as input (for example, transactions) must be the same of the output (for example, groups of transactions); however, in this case, the input are transactions and the output are rules. Thus, the metrics presented here are important as a final step to evaluate the usefulness of a clustering process in obtaining groups of rules. Therefore, the works described in this section are the ones that aim to apply clustering as a previous step to obtain association rules and their limitations regarding an assessment procedure.

3   Understanding the Issues Through a Subjective Evaluation

A subjective evaluation was done with some users to understand each issue before proposing the metrics. The analysis was done as a way to comprehend the important aspects to be considered by a user when comparing the performance of some methodologies or even their results.

The evaluation was done through a questionnaire, which was elaborated to present to the users scenarios that can occur in the considered context. For each scenario a question was formulated in order to comprehend the importance of the aspects related to the exposed problem. The full questionnaire can be found in *Appendix*. Before presenting the Questions, a brief introduction was given to the users about the questionnaire. The questionnaire comprehends 10 Questions: Questions 1 and 2 are related to "Issue 1", Questions 3, 4, 5 and 6 to "Issue 2" and Question 7 to "Issue 3". Question 8 is related to a more general advantage the clustering process can provide. Questions 9 and 10 are general questions about the evaluation. The questionnaire was answered by 5 users, all of them researches on data mining and/or text mining. All of them have knowledge on clustering and association rules, including their particularities, problems, means of validation, etc., and, so, they could understand the exposed problem and contribute with the current research. A brief overview of their experiences is here presented: (user (1)) worked with objective measures to post-process association rules obtained to extract related words in a collection of documents to construct a enriched bag-of-words representation; worked with clustering to construct a topic hierarchy based on features extracted through association rules; (user (2)) worked with clustering to post-process association rules and developed a labeling method to label groups of rules; worked with complex networks and semi-supervised learning to post-process association rules; (user (3)) worked with clustering and semi-supervised learning to build a dynamic topic hierarchy; worked with association rules in some applications; (user (4)) worked with objective and subjective measures to post-process association rules; worked with clustering to organize document collections; (user (5)) worked with taxonomies to generalize association rules to help users during the post-processing phase; worked with association rules to develop recommendation systems; worked with clustering in some applications. The users' answers are presented in Table 1, which are discussed below along with each Question.

Table 1. Users' evaluation about the exposed scenarios.

Question	User				
	1	2	3	4	5
1	desirable	desirable	desirable	desirable	desirable
2	desirable	desirable	desirable	desirable	indifferent
3	desirable	desirable	indifferent : no	desirable	desirable
4	indifferent : no	indifferent	indifferent : no	desirable	desirable
5	indifferent : no	desirable	desirable	desirable	indifferent
6	desirable	no desirable	desirable	no desirable	desirable
7	low	high	low	average	average
8	yes	yes	yes	yes	yes

Question 1 (Issue 1). The idea behind this question, presented in Fig. 2, was to analyze if it is important not to lose knowledge during the clustering process. The other idea was to analyze, regarding the repetition over the clusters, if the knowledge must be in subsets that express its own associations. All of them considered that is *desirable* that rules extracted in RsT be found in RsP, directing the analysis to the aspects embedded in question. Besides, all of them stated that this aspect is important for an assessment procedure (1.a.) – the "yes" answers, in all the questions, regarding the importance of the considered aspect, are omitted in Table 1. In relation to the comments made by the users (1.b.), we can mention: (a) the necessity to differentiate the rules that don't repeat over the clusters from the ones that repeat; (b) the understanding that if the patterns are present in both sets it is because they are relevant (even if they, probably, represent a frequent relation of the domain), being desirable that this scenario occurs. To formalize these aspects, two metrics are proposed in the next section: M_{O-RsP} and $M_{R-O-RsP}$.

---

1. In your opinion, observing "Scenario-A" below, how do you consider the occurrence of rules obtained in RsT in RsP (cases in green and orange)? Both the cases, green and orange, represent rules obtained in both of the sets, but the rules in orange are extracted more than once in RsP along the groups. If needed to distinguish the green cases of the orange cases, please let it indicated.
( ) desirable ( ) indifferent ( ) no desirable

a. Do you think important to consider this scenario in an assessment procedure to be used in the presented context?
( ) yes ( ) no

b. Would you like to make any comment about the scenario (advantage, disadvantage, etc.)?

Fig. 2. Question 1 (see *Appendix* for details).

Question 2 (Issue 1). The idea behind this question, presented in Fig. 3, was to analyze if it is important to obtain new knowledge during the clustering process. The other idea was to analyze, regarding the repetition over the clusters, if the new knowledge must be in subsets that express its own associations. Almost all of them (four out five) considered that is *desirable* that rules extracted in RsP be not found in RsT, directing the analysis to the aspects embedded in question. However, all of them stated that this aspect is important for an assessment procedure (2.a.). In relation to the comments made by the users (2.b.), we can mention: (a) the necessity to differentiate the rules that don't repeat over the clusters from the ones that repeat; (b) the understanding that the patterns appearing only in RsP represent new knowledge, probably interesting to the users, that would be difficult to be found through the traditional process. To formalize these aspects, two metrics are proposed in the next section: M_{N-RsP} and $M_{R-N-RsP}$.

---

2. In your opinion, observing "Scenario-A" below, how do you consider the non-occurrence of rules obtained in RsP in RsT (cases in purple and red)? Both the cases, purple and red, represent rules obtained only in RsP, but the rules in red are extracted more than once in RsP along the groups. If needed to distinguish the purple cases of the red cases, please let it indicated.
( ) desirable ( ) indifferent ( ) no desirable

a. Do you think important to consider this scenario in an assessment procedure to be used in the presented context?
( ) yes ( ) no

b. Would you like to make any comment about the scenario (advantage, disadvantage, etc.)?

Fig. 3. Question 2 (see *Appendix* for details).

Question 3 (Issue 2). The idea behind this question, presented in Fig. 4, was to analyze if it is important to discovery new *interesting* knowledge; if so, the cost of the clustering process would be minimized since new interesting knowledge would be found. Almost all of them (four out five) considered that is *desirable* that some (or none) of the n most interesting rules in RsP be not found in RsT, directing the analysis to the aspect embedded in question. Besides, almost all of them (four out five) stated that this aspect is important for an assessment procedure (3.a.). In relation to the comments made by the users (3.b.), we can mention: (a) the importance of identifying new interesting rules not extracted through the traditional process; (b) the understanding that these new interesting patterns are the ones that stand in a certain group of data. To formalize this aspect, one metric is proposed in the next section: $M_{N-I-RsP}$.

Question 4 (Issue 2). The idea behind this question, presented in Fig. 5, was to analyze if it is important not to lose interesting knowledge during the clustering process. Most of them (three out five) considered that is *indifferent* that some (or none) of the n most interesting rules in RsT be not found in RsP; thus, in this case, any conclusion was done in relation to the aspect embedded in question. However, most of them (three out five) stated that

---

3. In your opinion, observing "Scenario-B" below, how do you consider the non-occurrence of some (or none) of the *n* more interesting rules in RsP in RsT (cases in blue)? Notice that the blue rules belong only to the RsP set.

( ) desirable ( ) indifferent ( ) no desirable

a. Do you think important to consider this scenario in an assessment procedure to be used in the presented context?

( ) yes ( ) no

b. Would you like to make any comment about the scenario (advantage, disadvantage, etc.)?

Fig. 4. Question 3 (see *Appendix* for details).

this aspect is important for an assessment procedure (4.a.). In relation to the comments made by the users (4.b.), there were few relevant observations. Only one user mentioned that it is important to analyze the considered scenario because it is understood that the clustering process could be losing knowledge, making the user to think about the truth quality of the clusters (directing to the aspect embedded in question). Although no conclusion could be made in relation to the main question, considering the importance for an assessment procedure (4.a.), one metric is proposed in the next section to formalize the aspect: $M_{O-I-N-RsP}$.

---

4. In your opinion, observing "Scenario-B" below, how do you consider the reverse scenario? This is, the non-occurrence of some (or none) of the *n* more interesting rules in RsT in RsP (cases in orange)? Notice that the orange rules belong only to the RsT set.

( ) desirable ( ) indifferent ( ) no desirable

a. Do you think important to consider this scenario in an assessment procedure to be used in the presented context?

( ) yes ( ) no

b. Would you like to make any comment about the scenario (advantage, disadvantage, etc.)?

Fig. 5. Question 4 (see *Appendix* for details).

Question 5 (Issue 2). The idea behind this question, presented in Fig. 6, was to analyze if it is important that an intersection exists between the new interesting knowledge, discovery through the clustering process, in relation to the interesting knowledge already found through the traditional process. Most of them (three out five) considered as *desirable* the existing intersection between the *n* most interesting rules in RsP and the *n* most interesting rules in RsT, directing the analysis to the aspect embedded in question. Besides, almost all of them (four out five) stated that this aspect is important for an assessment procedure (5.a.). In relation to the comments made by the users (5.b.), we can mention: (a) the understanding that interesting rules appearing in both sets have a high probability of being really interesting to the users; (b) however, as an opposed idea, the understanding that different knowledge must be extracted from the clustering process compared to the traditional one (*indifferent* cases). Thus, although distinct views exist (5.b.), considering the importance for an assessment procedure (5.a.), one metric is proposed in the next section to formalize the aspect: M_{C-I} (which can be interpreted according to the user's view).

---

5. In your opinion, observing "Scenario-B" below, how do you consider the existing intersection between the *n* more interesting rules in RsP and the *n* more interesting rules in RsT (cases in red)?
( ) desirable ( ) indifferent ( ) no desirable

a. Do you think important to consider this scenario in an assessment procedure to be used in the presented context?
( ) yes ( ) no

b. Would you like to make any comment about the scenario (advantage, disadvantage, etc.)?

Fig. 6. Question 5 (see *Appendix* for details).

Question 6 (Issue 2). The idea behind this question, presented in Fig. 7, was to analyze if it is important that the new interesting knowledge be found in a small number of the groups. Most of them (three out five) considered as *desirable* the spread of the *n* most interesting rules in RsP in a small number of clusters, directing the analysis to the aspect embedded in question. Besides, all of them stated that this aspect is important for an assessment procedure (6.a.). In relation to the comments made by the users (6.b.), we can mention: (a) the understanding that the user could explore a group of interesting knowledge, directing the user to a reduced exploration space; (b) however, as an opposed idea, the understanding that it could be more relevant to explore the interesting rules within each cluster, since each group express different concepts of the domain (*no desirable* cases). Thus, although distinct views exist (6.b.), considering the importance for an assessment procedure (6.a.), one metric is proposed in the next section to formalize the aspect: $M_{NC-I-RsP}$ (which can be interpreted according to the user's view).

---

6. In your opinion, how do you would consider the spread of the *n* more interesting rules in RsP in a small number of clusters?
( ) desirable ( ) indifferent ( ) no desirable

a. Do you think important to consider this scenario in an assessment procedure to be used in the presented context?
( ) yes ( ) no

b. Would you like to make any comment about the scenario (advantage, disadvantage, etc.)?

Fig. 7. Question 6 (see *Appendix* for details).

Question 7 (Issue 3). The idea behind this question, presented in Fig. 8, was to analyze the clustering process regarding the number of extracted rules in relation to the traditional one. *Two* of them considered that the amount of rules to be extracted through clustering, compared to the traditional process, should be *low*, *two* of them *average* and *one high*. However, all of them stated that this aspect is important for an assessment procedure (7.a.). In relation to the comments made by the users (7.b.), we can mention: (a) the understanding that, since each group contains a specific bunch of items, the amount of rules should be low (*low* cases); (b) the understanding that each group will contain a small number of rules (due to the same arguments of (a)) and, as a consequence, an average amount considering all the groups (*average* cases); (c) the understanding that since each group will contain similar items, the amount of rules should be high (due to the low diversity of items compared to the traditional process considering the same value of support/confidence) (*high* cases). Thus, although distinct views exist (7.b.), considering the importance for an assessment procedure (7.a.), one metric is proposed in the next section to formalize the aspect: M_{NR-RsP} (which can be interpreted according to the user's view).

---

7. In your opinion, do you consider that the amount of rules to be extracted through clustering, compared to the traditional process, should be:
( ) low ( ) average ( ) high

a. Do you think important to consider this scenario in an assessment procedure to be used in the presented context?
( ) yes ( ) no

b. Would you like to make any comment about the scenario (advantage, disadvantage, etc.)?

Fig. 8. Question 7.

Question 8 (Others). The idea behind this question, presented in Fig. 9, was to analyze the clustering process regarding the spread of the concepts of the domain. All of them considered that is *desirable* that the clustering process should, as a consequence, enable each cluster to express a distinct topic of the domain. Besides, all of them stated that this aspect is important for an assessment procedure (8.a.). In relation to the comments made by the users (8.b.), we can mention, unanimously, the understanding that the aim of the clustering process is to split the distinct topics of the domain in order to aid rule exploration. To formalize this aspect, two metrics are proposed in the next section: M_P and M_{RF}.

---

8. In your opinion, only in relation to RsP, do you consider that the clustering process should, as a consequence, enable each cluster to express a distinct topic of the domain?
( ) yes ( ) indifferent ( ) no

a. Do you think important to consider this scenario in an assessment procedure to be used in the presented context?
( ) yes ( ) no

b. Would you like to make any comment about the scenario (advantage, disadvantage, etc.)?

Fig. 9. Question 8.

Question 9 and 10 (Others). Questions 9 and 10 are general questions about the evaluation. In relation to "Question 9", presented in Fig. 10, none of them gave any suggestion. However, in relation to "Question 10", also presented in Fig. 10, two comments were made: (a) the understanding that an assessment procedure should consider two or more scenarios but maintaining a good trade-off among them; (b) the understanding that the presentation of the rules in different scenarios, as the ones exposed, can aid the user during an evaluation process, since the user can give more attention to some aspects according to his aims. Therefore, as observed, this paper contributes with current researches making the users think about some aspects to be considered in the presented context and providing to them a flexible way to explore their problems (since the users can interpret a metric according to their view).

---

9. Can you identify other scenario(s), not previously explored, that can be relevant to the presented context? Give an example of the scenario(s) that you identified.

   a. Do you think important to consider this(these) scenario(s) in an assessment procedure to be used in the presented context?
   ( ) yes ( ) no

10.If you want to leave any comment/observation, please do it below.

Fig. 10. Questions 9 and 10.

4   Evaluation Metrics: Providing an Assessment Procedure

Eleven metrics are proposed to provide an assessment procedure in order to support the evaluation of the methodologies that use clustering in the pre-processing step (as the ones described in Sect. 2). These metrics formalize the aspects related to each issue, which were analyzed by some users through a subjective evaluation. Each metric, which was implicit explored through a subjective question, is related to an issue. For each issue there are one or more metrics. All metrics, with exception to M_{NR-RsP}, range from 0 to 1. Since RsP contains all the rules extracted within each group, repeated rules may exist in the set; therefore, RsP can be, in some cases, a multiset. In RsT the same doesn't occur since the rules are unique.

Issue 1. Regarding the existing overlap among the rules in RsP and RsT, four metrics are proposed, which are described as follows:

M_{O-RsP} Related to "Question 1", measures the ratio of "old" rules in RsP, i.e., the ratio of rules in RsT found in RsP (Eq. 6). A rule is considered "old" if it is in RsT, i.e., in the rule set obtained through the traditional process. Therefore, considering users' analysis, the higher the value the better the metric, since the value indicates that there was no loss of knowledge during the process.

$$M_{O-RsP} = \frac{|RsT \cap RsP|}{|RsT|}. \tag{6}$$

$M_{R-O-RsP}$ Related to "Question 1", measures the ratio of "old" rules that repeat in RsP (Eq. 7). In this work, it is assumed that a rule should exist in only one of the clustering groups, since it has to be in a subdomain that expresses its own associations. Therefore, considering users' analysis and authors' understanding on the problem, the lower the value the better the metric, since the value indicates that the knowledge, already known, is in subsets that express its own associations.

$$M_{R-O-RsP} = \frac{FindRepetitionRsP(RsT \cap RsP)}{|RsT \cap RsP|}, \tag{7}$$

FindRepetitionRsP: function that receives by parameter a set of non repeated rules and returns the number of rules in the set that repeat in RsP.

M_{N-RsP} Related to "Question 2", measures the ratio of "new" rules in RsP, i.e., the ratio of rules in RsP not found in RsT (Eq. 8). A rule is "new" if it isn't in RsT, i.e., in the rule set obtained through the traditional process. Although it is important that any knowledge be lost (metric M_{O-RsP}), it is expected that the ratio of "new" rules in RsP be greater than the ratio of "old" rules. Therefore, considering users' analysis, the higher the value the better the metric, since the value indicates the amount of knowledge, previously unknown, obtained during the process.

$$M_{N-RsP} = \frac{|RsP^* - RsT|}{|RsP|}, \tag{8}$$

RsP*: all rules in RsP found in RsT are considered "old" rules and, therefore, are computed as having only one occurence in RsP.

$M_{R-N-RsP}$ Related to "Question 2", measures the ratio of "new" rules that repeat in RsP (Eq. 9). Idem to $M_{R-O-RsP}$. Therefore, considering users' analysis and authors' understanding on the problem, as in $M_{R-O-RsP}$, the lower the value the better the metric, since the value indicates that the knowledge, previously unknown, is in subsets that express its own associations.

$$M_{R-N-RsP} = \frac{FindRepetition(RsP^* - RsT)}{|RsP^* - RsT|}, \tag{9}$$

FindRepetition: function that receives by parameter a set that may contain repeated rules and returns the number of rules in the set that repeat; RsP*: idem Eq. 8.

Issue 2. Regarding the existing overlap among the rules in RsP and RsT considering the interesting aspect of the knowledge, four metrics are proposed, which are described as follows:

$M_{N-I-RsP}$ Related to "Question 3", measures the ratio of "new" rules among the h-top interesting rules in RsP (Eq. 10). Given a subset of h-top interesting rules, selected from RsP, it is expected that the ratio of "new" rules in this subset be as large as possible. The h-top rules are the h rules that contain the highest values regarding an objective measure, where h is a number to be chosen[2]. Therefore, considering users' analysis and authors' understanding on the problem, the higher the value the better the metric, since the value indicates that the cost of the process is minimized by the discovery of interesting knowledge, previously unknown, in RsP.

$$M_{N-I-RsP} = \frac{CountTopRules(h_{top} \ of \ RsP, RsP^* - RsT)}{|h_{top} \ of \ RsP|},$$

CountTopRules: function that receives by parameter a set of h-top interesting rules and a set of rules and returns the number of rules that appears among the h-top; RsP*: idem Eq. 8.

(10)

$M_{O-I-N-RsP}$ Related to "Question 4", measures the ratio of "old" rules not in RsP among the h-top interesting rules in RsT (Eq. 11). Given a subset of h-top interesting rules, selected from RsT, it is expected that all these rules are present in RsP. Since any conclusion could be done considering users' analysis, it is understood, in this work, that is not desirable that patterns in RsT disappear in RsP, which would imply in the loss of relevant knowledge. Thus, this metric measures the ratio of "old" interesting rules not in RsP. The h-top rules are as described in $M_{N-I-RsP}$. Therefore, considering the importance of this aspect for an assessment procedure, according to users' view, and authors' understanding on the problem, the lower the value the better the metric, since the value indicates that the interesting knowledge in RsT was not lost during the process.

$$M_{O-I-N-RsP} = \frac{CountTopRules(h_{top} \ of \ RsT, RsT - RsP)}{|h_{top} \ of \ RsT|},$$

CountTopRules: idem Eq. 10.

(11)

M_{C-I} Related to "Question 5", measures the ratio of common rules among the h-top interesting rules in RsP and the h-top interesting rules in RsT (Eq. 12). Consider two subsets, S_1 and S_2, containing, respectively, the h-top interesting rules in RsP and the h-top interesting rules in RsT. This metric measures the existing intersection between these two subsets, which is expected to be as large as possible according to the users (different from what was assumed in [12], that presented a metric interpretation as in the *indifferent* cases (see the justification below)). Therefore, the higher the value the better the metric, since the users understand that rules present in both sets have a high probability of being really interesting to them. However, the value of the metric can be interpreted according to the user's view, being the lower the value the better the metric (*indifferent* cases – in this condition,

---

[2] Any other criteria could be adopted to select the h-top interesting rules.

the process would not provide any additional relevant information, since all the knowledge already known as interesting in RsT is also identified as interesting in RsP, as understood in [12]). Therefore, as noticed, the metric provides a flexible way to analyze the problem.

$$M_{C-I} = \frac{|h_{top} \; of \; RsP \cap h_{top} \; of \; RsT|}{h},$$

h is the number to be chosen to realize the selection of the rules in both sets, i.e., RsP and RsT. $\qquad$ (12)

$M_{NC-I-RsP}$ Related to "Question 6", measures the ratio of groups in the clustering related to RsP that contains the h-top interesting rules in RsP (Eq. 13). Therefore, considering users' analysis, the lower the value the better the metric. This means that just some of the groups would have to be explored by the user, which will contain the "new" relevant knowledge extracted during the process. However, the value of the metric can be interpreted according to the user's view, being the higher the value the better the metric (*indifferent* cases – in this condition, each cluster would express its own interesting knowledge). As noticed, the metric provides a flexible way to analyze the problem.

$$M_{NC-I-RsP} = \frac{FindGroups(h_{top} \; of \; RsP)}{N},$$

N: number of groups in the clustering; FindGroups: function that receives by parameter a set of h-top interesting rules, finds their groups and returns the number of distinct selected groups. $\qquad$ (13)

Issue 3. Regarding the process behavior related to the number of rules that are obtained in RsP, a unique metric is proposed, which is described as follows:

M_{NR-RsP} Related to "Question 7", measures the ratio of rules in RsP in relation to RsT (Eq. 14). It is important to analyze the process in relation to the number of rules in RsP. It is not desirable, according to the authors' understanding, to have a large increase in the volume of rules, because even if new patterns are discovered, it will be harder for the user to identify them. Therefore, the lower the value the better the metric, since the value indicates that although new patterns have been extracted, the number of extracted rules is not big enough to overload the user. However, the value of the metric can be interpreted according to the user's view, since each user has a different opinion (although all of them agree with the importance of this aspect for an assessment procedure). As noticed, the metric provides a flexible way to analyze the problem.

$$M_{NR-RsP} = \frac{|RsP|}{|RsT|}. \qquad (14)$$

Others. "Question 8" tried to capture a more general advantage the clustering process can provide: to enable each cluster expresses a distinct topic of the domain.

According to the users, this is a *desirable* aspect. Carvalho et al. [21] evaluated some labeling methods for association rule clustering through two measures. Precision (P) measures how much a labeling method finds labels that represent as accurately as possible the rules contained in their own groups – if the labels don't represent the knowledge inside each cluster the user will have difficult to explore the existing concepts related to the topics the labels express. Repetition Frequency (RF) measures how the labels are distributed over the clusters – if the labels appear repeatedly over the clusters the user will have difficult to identify the existing topics. As known, a labeling method is applied over the clusters obtained through a clustering process. Thus, considering that a good labeling method exists[3], it is assumed that a methodology, in the presented context, provides a good distribution of the topics if it presents high values for P and RF, since which will impact the results is the clustering itself. Therefore, these measures are used as metrics in the considered context, which are described as follows:

M_P Considering that a good labeling method is available, measures how much the labels, built over the obtained clustering, represent the rules contained in the clusters (Eq. 15). The more a methodology provides groups that express their own associations, more specific domain knowledge the groups will contain and, probably, the more the labels will represent the rules in the clusters. Therefore, the higher the value the better the metric, since the value indicates that the methodology succeed to enable a suitable distribution of the domain topics.

$$M_P = \frac{\sum_{i=1}^{N} P(C_i)}{N}, P(C_i) = \frac{\#\{rules\ covered\ in\ C_i\ by\ C_i\ labels\}}{\#\{rules\ in\ C_i\}}, \tag{15}$$

N refers to the number of clusters. A rule is covered (represented) by a set of labels if the rule contains at least one of the labels.

M_{RF} Considering that a good labeling method is available, measures how much the distinct labels, built over the obtained clustering, don't repeat (Eq. 16). The more a methodology provides groups that express their own associations, more specific domain knowledge the groups will contain and, probably, the more distinct the labels will be over the clusters. Therefore, the higher the value the better the metric, since the value indicates that the methodology succeed to enable a suitable distribution of the domain topics.

$$M_{RF} = 1 - \frac{\#\{distinct\ labels\ that\ repeat\ in\ the\ clusters\}}{\#\{distinct\ labels\ in\ the\ clusters\}}. \tag{16}$$

Table 2 summarizes the metrics above described, indicating the suitable use of each one. As noticed, the metrics provide a flexible way to analyze the problem, since the user can analyze the results according to his interests (users can disagree about some aspects), both in relation to the metrics values' interpretation as in relation to the importance (weight) of each one to a given domain – as in other contexts, such as objective measures for association rules [23], where the

---

[3] In this work, it is considered that this labeling method is the one presented by [22].

Table 2. Summary and recommended use of the proposed evaluation metrics.

Metric	Description
$M_{O-RsP}[\uparrow]$	Ratio of "old" rules in RsP
	Recommended use: *to measure the loss of knowledge obtained by the clustering process*
$M_{R-O-RsP}[\downarrow]$	Ratio of "old" rules in RsP that repeat
	Recommended use: *to measure if the knowledge, already known, is in subsets that express their own associations*
$M_{N-RsP}[\uparrow]$	Ratio of "new" rules in RsP
	Recommended use: *to measure the amount of knowledge, previously unknown, obtained by the clustering process*
$M_{R-N-RsP}[\downarrow]$	Ratio of "new" rules in RsP that repeat
	Recommended use: *to measure if the knowledge, previously unknown, is in subsets that express their own associations*
$M_{N-I-RsP}[\uparrow]$	Ratio of "new" rules among the h-top interesting rules in RsP
	Recommended use: *to measure the amount of previously unknown patters among the knowledge, obtained by the clustering process, identified as interesting*
$M_{O-I-N-RsP}[\downarrow]$	Ratio of "old" rules not in RsP among the h-top interesting rules in RsT
	Recommended use: *to measure the amount of knowledge already known as interesting among the knowledge obtained by the clustering process*
$M_{C-I}[\uparrow]$	Ratio of common rules among the h-top interesting rules in RsP and the h-top interesting rules in RsT
	Recommended use: *to measure the amount of common interesting knowledge between the patterns already known as interesting and the patterns previously unknown that were identified as interesting*
$M_{NC-I-RsP}[\downarrow]$	Ratio of groups in the clustering related to RsP that contains the h-top interesting rules in RsP
	Recommended use: *to measure if the patterns, previously unknown, identified as interesting, are spread over a small number of groups*
$M_{NR-RsP}[\downarrow]$	Ratio of rules in RsP in relation to RsT
	Recommended use: *to measure if the number of patterns obtained by the clustering process don't overload the user*
$M_P[\uparrow]$	Behavior of topics' distribution regarding *precision*
	Recommended use: *to measure if an obtained clustering enables a suitable distribution of the domain topics regarding precision*
$M_{RF}[\uparrow]$	Behavior of topics' distribution regarding *repetition frequency*
	Recommended use: *to measure if an obtained clustering enables a suitable distribution of the domain topics regarding repetition frequency*

user chooses the measures (and sometimes their thresholds cut) to evaluate the obtained patterns according to his interests. Besides, the metrics also contribute in the sense of enabling the users to think about important aspects related to the presented context. Finally, relating the proposed metrics by the researchers found in the literature (Sect. 2), it can be observed that: (a) [7] is the only work that provides a similar analysis related to the metrics M_{O-RsP} and M_{N-RsP} in *Issue 1*; (b) none of them provide an analysis related to the aspects covered by *Issue 2* and *Others*; [7,11,13] provide a similar analysis related to the metric M_{NR-RsP} in *Issue 3*. Thus, as mentioned in Sect. 2, the necessity of an assessment procedure becomes evident.

5   Experiments

Some experiments were carried out in order to present how the metrics can be used. Therefore, two contexts were defined. Suppose a user decides to apply clustering in the pre-processing step. First of all, he has to find the most suitable methodology to use in his problem. After that, he has to check if the selected methodology was good enough for the problem, considering that different interests may be important for his decision. Thus, two different situations were regarded: (i) identify among some clustering setups the most suitable; (ii) analyze the process itself. A clustering setup is obtained by the application of a clustering algorithm combined with a similarity measure. Therefore, the metrics provide the support to evaluate each situation under the discussed issues: while in (i) the data is initially clustered through some clustering setups in order to identify the one that obtains a good association set, in (ii) the usefulness of the process itself is analyzed. Four data sets and four clustering setups were selected to be used in the experiments. It is important to mention that the choices below could be changed without affecting the paper relevance, since the aim here is to present the metrics and to demonstrate how they can be used, independently of the clustering setup.

The four data sets were Adult (48842;115), Income (6876;50), Groceries (9835;169) and Sup (1716;1939). The numbers in parenthesis indicate, respectively, the number of transactions and the number of distinct items in each data set. The first three are available in the R Project for Statistical Computing through the package "arules"[4]. The last one was donated by a supermarket located in São Carlos city, Brazil. Adult and Income are relational sets and Groceries and Sup transactional. Therefore, before extracting the rules on Adult and Income, the sets were converted to a transactional format, where each transaction was composed by pairs of the form "attribute=value".

The four clustering setups were obtained by the combination of the algorithms and similarity measures presented in Table 3. Each combination gives a clustering setup, i.e., a different way to analyze the process. PAM is a partitional medoid-based algorithm that splits n objects in k groups, in which each object is closer to the medoid that defines its own cluster than to the medoid of any

---

[4] http://cran.r-project.org/web/packages/arules/index.html.

other cluster – a medoid is an object that has the minimal average dissimilarity to all the other objects in the same cluster [24]. Therefore, the algorithm works searching for the k representative objects in n to built the k clusters by assigning each object to its nearest medoid. On the other hand, Ward is an agglomerative hierarchical algorithm that starts by considering each object as a singleton cluster and then, at each iteration, merges the two closest clusters until a single cluster remains. The two closest clusters are the ones that minimize the increase of the within-cluster sum of the squared errors (SSE) [24]. Despite the existence of algorithms designed for transactions, such as ROCK, the choices of the algorithms were made based on works that cluster the rules in the post-processing phase, as [25,26], aiming a *posteriori* comparison. The similarity measures were chosen considering the works described in Sect. 2 – only the similarities among transactions were selected based on previous experiments. Finally, although it is necessary to set k, the number of groups, in order to obtain a clustering setup, this value was only used to analyze the clustering setups on different views. We understand that even though k is an important parameter, its values were ranged and then averaged (see Sect. 6) without affecting the experiments' results, since the aim of the paper, as mentioned before, is to present some metrics and to demonstrate how they can be used, independently of the clustering setup. Besides, as it will be seen in Sect. 6, almost all the metrics present a low standard deviation, showing a good homogeneity among the results obtained over the values of k[5].

As described before, the rules are extracted within each group after clustering the data. The values of the minimum support (min-sup) and minimum confidence (min-conf) have to be set in order to extract a set of association rules. To automate the specification of the min-sup in each group, the following procedure was adopted: (i) find the 1-itemsets of the group with their supports, (ii) compute the average of these supports, (iii) use this average support as the min-sup of the group. This strategy was adopted since some groups can have few transactions and others many transactions, which implies in choosing suitable parameter values to avoid an explosion of rules within a group or the obtainment of an empty rule set. Regarding min-conf, the following values were used for each data set: Adult 50 %; Income 50 %; Groceries 10 %; Sup 100 %. Thus, the same min-conf value was applied in all the groups of a given data set. These values were chosen experimentally. Although it is known that min-sup and min-conf impact on the set of rules that are obtained, it was assumed that the focus was on the use of the metrics and, so, that the values were adequate for the proposed problem (the same argument is also applied to algorithms, similarity measures and k). The rules were extracted with an Apriori implementation developed by Christian Borgelt[6] with a minimum of two items and a maximum of five items per rule.

---

[5] In this work, each dendrogram obtained by Ward were cut considering each one of the values of k.

[6] http://www.borgelt.net/apriori.html.

Considering the four clustering setups, the RsP sets were obtained. However, once almost all the metrics are based on the rules obtained through the traditional process, the four data sets were also processed to obtain the RsT sets. For that, the min-sup was set automatically, as described before. Regarding min-conf, the same values used in RsP were considered, i.e., Adult 50 %, Income 50 %, Groceries 10 % and Sup 100 %. Furthermore, as some of the metrics are based on the h-top interesting rules of a given rule set, an objective measure should be selected. Instead of choosing a specific measure, the average rating obtained through 18 objective measures (see Table 3) was considered as follows: (i) the value of 18 measures was computed for each rule; (ii) each rule received 18 ID's, each ID corresponding to the rule position in one of the ranks related to a measure; (iii) the average was then calculated based on the rank positions (ID's). Thus, the h-top rules were selected considering the best average ratings. h, also a number to be set, was defined, in all the sets (RsT and RsP), to assume 10 % of the total of rules in RsT (always the smallest set). Therefore, each rule set contains its own values that are proportional in all of them.

To finish, as mentioned before, the labeling method applied over the clusters of each obtained clustering was the one presented by [22], named GLM (*G*enetic *L*abeling *M*ethod). In GLM the labels of each cluster are chosen by optimizing Precision (P) and Repetition Frequency (RF), the two measures previous described in Sect. 4. In fact, GLM is a genetic algorithm approach that aims to ensure a good tradeoff between P and RF. The fitness function of an individual is defined by $Fitness(I) = (P + RF) - \left(\frac{Max(P,RF)}{Min(P,RF)} * 10^{-5} \right)$, where (P+RF) shows how good are the measures according to the labels and $\left(\frac{Max(P,RF)}{Min(P,RF)} * 10^{-5} \right)$ the penalty proportional to the distance between P and RF. Table 3 summarizes the configurations used to apply the proposed metrics.

Table 3. Configurations used to apply the proposed metrics.

Data sets	Adult; Income; Groceries; Sup
Algorithms	PAM; Ward [algorithms details in [24]]
Similarity measures	Agg; Denza
k	5 to 25, steps of 5
h	10 % of the total of rules in RsT
Objective measures [measures details in [23]]	Added Value, Certainty Factor, Collective Strength, Confidence, Conviction, IS, ϕ-coefficient, Gini Index, J-Measure, Kappa, Klosgen, λ, Laplace, Lift, Mutual Information (asymmetric), Novelty, Support, Odds Ratio
Labeling Method	GLM [details in [22]]

6   Results and Discussion

Considering the configurations presented in Table 3 and the RsT sets above described, the experiments were carried out and the values of each metric obtained. Regarding the first proposed situation, i.e., identify among some clustering setups the most suitable (Sect. 5), an analysis based on the average of each metric was carried out. Table 4 presents the results for each data set. In this case, the metrics will help the users to find a suitable methodology for their problems. In order to aid the comparison of the results, all the metrics that present better results when their values are the smallest ($M_{R-O-RsP}$, for example) were processed to store the complement of the information. Therefore, all the metric, with exception to M_{NR-RsP}, have the same interpretation: the higher the value the better the performance. Furthermore, all the metrics can be seen in terms of percentage if multiplied by 100 (ex.: 0.807*100 = 80.7 %).

Each average in Table 4 was obtained from the results of the experiments related to the presented configuration. The value 0.807 in M_{O-RsP} at Adult:-PAM:Agg, for example, was obtained by the average of the values in M_{O-RsP} at Adult:PAM:Agg over the values of k. The table also presents, between "[ ]", the standard deviations; regarding the last example, the standard deviation is ±0.02 (0.807 [±0.02]). It can be observed that almost all the metrics present a low standard deviation, showing a good homogeneity among the results obtained over the values of k. The major exception is M_{NR-RsP}, which presents high standard deviation values, since the higher the number k of groups the higher the number of rules. For each data set, the highest averages are marked with ▲ in each metric. The only exception is M_{NR-RsP}, where the lowest averages are highlighted. For Adult, for example, the best average for $M_{R-O-RsP}$ is the one related to PAM:Agg (0.807). Thereby, it is possible to visualize, for each data set, the most suitable clustering setup. It is important to mention that the results are deterministic and, therefore, no statistical test was done to check if there is a significant difference among the averages. It can be noticed that:

Adult. The most suitable configuration for this data set is ***PAM:Agg***, since it presents better results in 8 of the 11 metrics. Furthermore, it can be noticed that in some cases the values in PAM:Agg are more representative than the others – observe, for example, that while in PAM:Agg M_{O-RsP} presents a performance above 80 %, the others presents a performance below 60 % (see also $M_{O-I-N-RsP}$ and M_{NR-RsP}).

Income. The most suitable configuration for this data set is ***PAM:Denza***, since it presents better results in 4 of the 11 metrics. However, PAM:Agg can also be useful, since it presents better results in 3 of the 11 metrics and the two setups present close values in almost all the metrics. Thus, in this case, the user can choose one of them based on the importance each metric represents to him in the considered problem.

Groceries. The most suitable configuration for this data set is ***Ward:Agg***, since it presents better results in 7 of the 11 metrics (in 5 excluding the ties). Furthermore, it can be noticed that in some cases the values in Ward:Agg are more

Table 4. Average of the proposed metrics, for each data set, in the considered clustering setups.

Adult

Algorithm	Measure	M_{O-RsP}	$M_{R-O-RsP}$	M_{N-RsP}	$M_{R-N-RsP}$	$M_{N-I-RsP}$	$M_{O-I-N-RsP}$
PAM	Agg	0.807 [±0.02]▲	0.343 [±0.07]▲	0.834 [±0.02]	0.890 [±0.01]▲	0.588 [±0.02]	0.840 [±0.01]▲
	Denza	0.585 [±0.03]	0.314 [±0.06]	0.888 [±0.03]▲	0.863 [±0.03]	0.824 [±0.03]	0.333 [±0.05]
	Measure	M_{C-I}	$M_{NC-I-RsP}$	M_{NR-RsP}	M_P	M_{RF}	
	Agg	0.778 [±0.02]	0.551 [±0.08]▲	12.221 [±2.53]▲	0.699 [±0.03]▲	0.830 [±0.06]▲	
	Denza	0.886 [±0.03]	0.483 [±0.15]	20.460 [±9.63]	0.657 [±0.03]	0.789 [±0.12]	
Ward	Agg	0.565 [±0.01]	0.253 [±0.07]	0.867 [±0.03]	0.854 [±0.02]	0.932 [±0.02]▲	0.150 [±0.01]
	Denza	0.565 [±0.04]	0.298 [±0.03]	0.878 [±0.03]	0.854 [±0.03]	0.870 [±0.04]	0.237 [±0.10]
	Measure	M_{C-I}	$M_{NC-I-RsP}$	M_{NR-RsP}	M_P	M_{RF}	
	Agg	0.993 [±0.01]▲	0.394 [±0.11]	17.587 [±7.99]	0.619 [±0.04]	0.805 [±0.11]	
	Denza	0.921 [±0.05]	0.385 [±0.22]	18.057 [±7.61]	0.618 [±0.04]	0.804 [±0.12]	

Income

Algorithm	Measure	M_{O-RsP}	$M_{R-O-RsP}$	M_{N-RsP}	$M_{R-N-RsP}$	$M_{N-I-RsP}$	$M_{O-I-N-RsP}$
PAM	Agg	0.909 [±0.03]▲	0.134 [±0.05]	0.979 [±0.01]	0.871 [±0.04]	0.844 [±0.10]	0.942 [±0.04]▲
	Denza	0.876 [±0.03]	0.156 [±0.10]	0.983 [±0.00]▲	0.894 [±0.03]	0.924 [±0.05]▲	0.833 [±0.08]
	Measure	M_{C-I}	$M_{NC-I-RsP}$	M_{NR-RsP}	M_P	M_{RF}	
	Agg	0.898 [±0.07]	0.640 [±0.15]	238.519 [±120.36]▲	0.654 [±0.06]	0.759 [±0.16]	
	Denza	0.938 [±0.03]	0.643 [±0.23]	300.608 [±169.34]	0.666 [±0.06]▲	0.786 [±0.13]▲	
Ward	Agg	0.871 [±0.07]	0.173 [±0.09]▲	0.980 [±0.00]	0.888 [±0.04]	0.909 [±0.05]	0.782 [±0.10]
	Denza	0.879 [±0.06]	0.120 [±0.10]	0.978 [±0.01]	0.901 [±0.04]▲	0.869 [±0.08]	0.782 [±0.08]
	Measure	M_{C-I}	$M_{NC-I-RsP}$	M_{NR-RsP}	M_P	M_{RF}	
	Agg	0.942 [±0.05]▲	0.612 [±0.12]	241.775 [±134.24]	0.632 [±0.07]	0.773 [±0.13]	
	Denza	0.916 [±0.04]	0.681 [±0.15]▲	249.930 [±139.44]	0.644 [±0.08]	0.773 [±0.11]	

Groceries

Algorithm	Measure	M_{O-RsP}	$M_{R-O-RsP}$	M_{N-RsP}	$M_{R-N-RsP}$	$M_{N-I-RsP}$	$M_{O-I-N-RsP}$
PAM	Agg	0.981 [±0.03]	0.429 [±0.17]	0.807 [±0.05]▲	0.948 [±0.03]	0.500 [±0.11]	1.000 [±0.00]▲
	Denza	1.000 [±0.00]▲	0.100 [±0.07]	0.798 [±0.07]	0.880 [±0.02]	0.567 [±0.17]▲	1.000 [±0.00]▲
	Measure	M_{C-I}	$M_{NC-I-RsP}$	M_{NR-RsP}	M_P	M_{RF}	
	Agg	1.000 [±0.00]▲	0.909 [±0.06]▲	11.238 [±4.41]	0.613 [±0.10]	0.833 [±0.10]	
	Denza	0.933 [±0.08]	0.714 [±0.19]	18.178 [±8.58]	0.558 [±0.10]	0.800 [±0.09]	
Ward	Agg	1.000 [±0.00]▲	0.894 [±0.13]▲	0.245 [±0.27]	0.997 [±0.01]▲	0.367 [±0.45]	1.000 [±0.00]▲
	Denza	1.000 [±0.00]▲	0.509 [±0.37]	0.538 [±0.27]	0.949 [±0.06]	0.567 [±0.36]▲	1.000 [±0.00]▲
	Measure	M_{C-I}	$M_{NC-I-RsP}$	M_{NR-RsP}	M_P	M_{RF}	
	Agg	0.367 [±0.45]	0.792 [±0.09]	1.866 [±1.05]▲	0.893 [±0.09]▲	0.966 [±0.04]▲	
	Denza	0.700 [±0.40]	0.850 [±0.03]	5.978 [±4.29]	0.758 [±0.15]	0.950 [±0.08]	

Sup

Algorithm	Measure	M_{O-RsP}	$M_{R-O-RsP}$	M_{N-RsP}	$M_{R-N-RsP}$	$M_{N-I-RsP}$	$M_{O-I-N-RsP}$
PAM	Agg	0.778 [±0.03]	0.990 [±0.01]	0.996 [±0.00]▲	0.999 [±0.00]	1.000 [±0.00]▲	1.000 [±0.00]▲
	Denza	0.849 [±0.09]	0.920 [±0.09]	0.971 [±0.05]	0.999 [±0.00]	0.924 [±0.15]	0.972 [±0.03]
	Measure	M_{C-I}	$M_{NC-I-RsP}$	M_{NR-RsP}	M_P	M_{RF}	
	Agg	1.000 [±0.00]▲	0.909 [±0.06]▲	528.422 [±735.64]	0.496 [±0.06]	0.971 [±0.03]	
	Denza	0.959 [±0.08]	0.855 [±0.13]	845.059 [±942.35]	0.538 [±0.09]	0.960 [±0.03]	
Ward	Agg	0.952 [±0.05]▲	1.000 [±0.00]▲	0.245 [±0.25]	1.000 [±0.00]▲	0.055 [±0.07]	0.993 [±0.01]
	Denza	0.946 [±0.05]	1.000 [±0.00]▲	0.848 [±0.11]	0.999 [±0.00]	0.641 [±0.24]	0.993 [±0.01]
	Measure	M_{C-I}	$M_{NC-I-RsP}$	M_{NR-RsP}	M_P	M_{RF}	
	Agg	0.228 [±0.10]	0.864 [±0.04]	1.439 [±0.54]▲	0.961 [±0.02]▲	1.000 [±0.00]▲	
	Denza	0.669 [±0.20]	0.885 [±0.05]	437.856 [±863.97]	0.862 [±0.05]	0.978 [±0.03]	

representative than the others – observe, for example, that while in Ward:Agg $M_{R-O-RsP}$ presents a performance above 89 %, the others presents a performance below 50 % (see also M_{NR-RsP} and M_P).

Sup. The most suitable configuration for this data set is **Ward:Agg**, since it presents better results in 6 of the 11 metrics (in 5 excluding the ties). Furthermore, it can be noticed that in some cases the values in Ward:Agg are more representative than the others (see M_{NR-RsP} and M_P). However, PAM:Agg can also be useful, since it presents better results in 5 of the 11 metrics (also 5 excluding the ties). Thus, in this case, the user can choose one of them based on the importance each metric represents to him in the considered problem.

Therefore, considering the user selected and used the configurations presented in Table 3, it can be observed that the most suitable clustering setup according to the metrics is PAM:Agg for Adult, PAM:Denza for Income and Ward:Agg for Groceries and Sup. In other words, the user will obtain better results, i.e., reasonable rule sets, if he initially clusters Adult through PAM:Agg, Income through PAM:Denza and Groceries and Sup through Ward:Agg. However, in other situations, different aspects can be of interesting, providing to the user a flexible way to explore his problems: the user can choose the metrics to apply and their better interpretations (lower/higher values (as mentioned in some of the metrics (Sect. 4))). Thus, in this first situation, the metrics provide criteria to carry out comparisons, helping the user to select the most suitable methodology for his problem.

From that point, supposing that PAM:Agg is a suitable solution for the user's problem regarding Adult, it is possible to analyze the process itself, i.e., to check if good results are really obtained (the same is valid for the other data sets). Observe that different interests may be important for his decision. Thus, the metrics provide criteria not only to analyze the process, but also to identify its positive and negative aspects, helping the user to reach a conclusion. To discuss this second situation, Table 5 presents the values of the metrics, in the selected clustering setup, regarding Adult (although this second situation is only discussed on Adult, the same analysis can be done to the other data sets). These values are the ones presented in Table 4, but in their original scales, since the smaller scales ($\downarrow$) were previous converted – the larger scales ($\uparrow$) remain the same. The scale, in each metric, is found between "[ ]". It can be noticed that:

M_{O-RsP}: little knowledge is lost during the process, around 20 %, since more than 80 % of the rules in RsT are found in RsP – being a positive aspect.

$M_{R-O-RsP}$: the repetition of "old" rules in RsP is high, around 66 %, indicating that the knowledge, already known, is not in subdomains that express its own associations – being a negative aspect.

M_{N-RsP}: most of the rules in RsP are "new", around 83 %, indicating the discovery of a great amount of knowledge previously unknown – being a positive aspect.

$M_{R-N-RsP}$: the repetition of "new" rules in RsP is low, around 11 %, indicating that the knowledge, previously unknown, is in subdomains that express its own associations – being a positive aspect.

$M_{N-I-RsP}$: nearly half of the h-top interesting rules in RsP are "new", around 59 %, indicating that the cost of the process is minimized by the discovery of interesting knowledge, previously unknown, in RsP – being a positive aspect.

$M_{O-I-N-RsP}$: the loss of "old" and interesting knowledge is low, around 16 %, since a great amount of the h-top interesting rules in RsT are found in RsP, around 84 % (100 % – 16 %) – being a positive aspect.

M_{C-I}: the intersection between the h-top interesting rules in RsP and the h-top interesting rules in RsT is high, around 78 %, indicating that a great amount of the knowledge already known as interesting in RsT is found in RsP – being a positive aspect.

$M_{NC-I-RsP}$: the number of groups that contain the h-top interesting rules in RsP is high, around 45 % – being a negative aspect, since many groups would have to be explored once the "new" relevant knowledge of the domain would be spread over the clusters.

M_{NR-RsP}: the number of rules in RsP is only around 12 times greater in relation to RsT – being a positive aspect (it can be seen in Table 4 that this ratio can be very high).

M_P: the selected clustering setup provides a suitable distribution of the domain topics, since the labels represent the rules in the clusters at a ratio around 70 % – being a positive aspect.

M_{RE}: the selected clustering setup provides a suitable distribution of the domain topics, since the distinct labels don't repeat over the clusters at a ratio around 83 % – being a positive aspect.

Table 5. Average of the proposed metrics in Adult, PAM:Agg clustering setup.

Data set	M_{O-RsP} [↑]	$M_{R-O-RsP}$ [↓]	M_{N-RsP} [↑]	$M_{R-N-RsP}$ [↓]	$M_{N-I-RsP}$ [↑]	$M_{O-I-N-RsP}$ [↓]
	0.807	0.657	0.834	0.110	0.588	0.160
Adult	M_{C-I} [↑]	$M_{NC-I-RsP}$ [↓]	M_{NR-RsP} [↓]	M_P [↑]	M_{RF} [↑]	
PAM:Agg	0.778	0.449	12.221	0.699	0.830	

As seen, considering the positive and negative aspects of the process, the user can analyze the results, according to his interests, and conclude if good results were reached. It is relevant to mention that the importance of each percentage depends on the user's needs, on the data sets, etc., and, therefore, the metrics provide a flexible way for him to explore the problems. Regarding the presented context, it can be said that the process obtains reasonable results, since 9 of the 11 aspects were considered positives. However, if the weight of the negative aspects is more important to the user, he can discard the results. Moreover, he can try to improve the process to obtain better results in these metrics, since he has an overview of all the aspects. Thus, in this second situation, the metrics provide criteria to analyze the process based on different interests, identifying its positive and negative aspects, helping the user to reach a conclusion.

Finally, to complement and finalize the discussion, Fig. 11 presents the variation of h parameter in the four metrics that depend on that value: $M_{N-I-RsP}$, $M_{O-I-N-RsP}$, M_{C-I}, $M_{NC-I-RsP}$. The clustering setup related to each data set is the one above identified as the most suitable. Axis x is related to h and axis y to the metrics' values. The metrics are represented by the different lines in the graphics. Note that the metrics' values related to $h=10$ % are the same as the ones presented in Table 4 (as before, each metric's value was obtained by the average of the values in the metric over the values of k). It can be seen that:

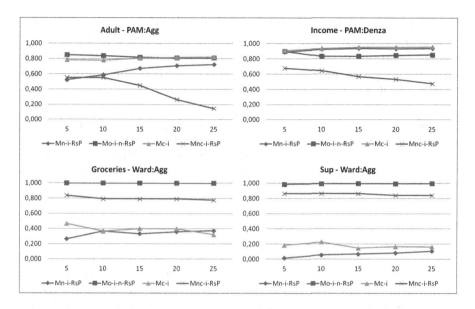

Fig. 11. Behavior of the h parameter in $M_{N-I-RsP}$, $M_{O-I-N-RsP}$, M_{C-I}, $M_{NC-I-RsP}$ in each data set (Color figure online).

(a) $M_{N-I-RsP}$ [↑] (blue line) tends to become higher as h increases or tends to assume a value close to the other h values; therefore, this metric presents better results with high values of h; (b) $M_{O-I-N-RsP}$ [↓] (red line) tends to decrease as h increases or tends to assume a value close to the other h values; therefore, this metric presents better results with high values of h; (c) M_{C-I} [↑] (green line) does not present a pattern; however, while in the relational data sets the values are close and high regardless the h value in the transactional ones the values are close and low regardless the h value; (d) $M_{NC-I-RsP}$ [↓] (purple line) tends to decrease as h increases or tends to assume a value close to the other h values; therefore, this metric presents better results with high values of h.

7 Conclusion

In this paper, eleven metrics were proposed to provide an assessment procedure in order to support the evaluation of methodologies that use clustering in the pre-processing step. The metrics were developed to answer three main issues. However, to propose the metrics, a subjective evaluation was done with some users to understand each issue. Some experiments were carried out in order to present how the metrics can be used. For that, two different situations were regarded: (i) identify among some clustering setups the most suitable; (ii) analyze the process itself. Through the experiments, it could be noticed that the metrics provide criteria to: (a) analyze the methodologies, (b) identify their positive and negative aspects, (c) carry out comparisons among them and, therefore,

(d) help the users to select the most suitable solution for their problems. Besides, the metrics do the users think about aspects to be considered in the presented context and provide to them a flexible way to explore the problems. Finally, this paper complements [12]'s work, since the subjective evaluation is used to attest the importance of the previous obtained results.

Acknowledgments. We wish to thank Fundação de Amparo à Pesquisa do Estado de São Paulo (FAPESP) (processes numbers: 2010/07879-0 and 2011/19850-9) and Coordenação de Aperfeiçoamento de Pessoal de Nível Superior (CAPES) (process number DS-6345378/D) for the financial support. Besides, we also want to thank the reviewers for the great contributions.

Appendix: Questionnaire

Introduction. Many issues related to association rule mining have received attention in the last years, especially the ones aiming to discover and facilitate the search for the interesting patterns of the domain. One approach related to this issue is the application of clustering in the pre-process step. In this case, as noticed in the figure below, data are initially grouped in n groups $(GD_1, GD_2, ..., GD_n)$. From this initial clustering, the rules are then extracted within each group (cluster), obtaining n groups of rules $(GR_1, GR_2, ..., GR_n)$. *The aim is to obtain potentially interesting rules that would not be extracted from unpartitioned data sets, for not having enough support, without overloading the user with a great amount of patterns.* The user must set the minimum support to a low value to discover these same patterns from unpartitioned data sets, causing a rapidly increase in the number of rules. Thereby, data are initially split and the rules are extracted within each group, in a manner that *each group expresses its own associations without the interference of the other groups that contain different association patterns*. Distinct methodologies have been proposed to enable this process. Each methodology uses a different combination of clustering algorithms and similarity measures in order to obtain the groups of rules.

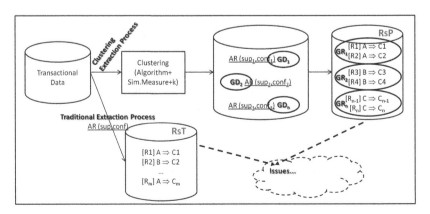

It is in this context that this evaluation should be done. Some scenarios that can occur in this scope are shown below, waiting for your contribution for a better understanding of the problem. In all the cases, it is assumed that two rule sets are available, in order to evaluate the presented scenarios: one extracted through traditional process, RsT[7], and one extracted through clustering (process above described), RsP[8] – the examples presented below are merely illustrations of the scenarios and, therefore, should not be evaluated considering the knowledge they express. Based on this evaluation, the aim is to propose an assessment procedure to support the analysis of the existing methodologies.

Scenarios

1. In your opinion, observing "Scenario-A" (Table 6), how do you consider the occurrence of rules obtained in RsT in RsP (cases in green and orange)? Both the cases, green and orange, represent rules obtained in both of the sets, but the rules in orange are extracted more than once in RsP over the groups. If needed to distinguish the green cases of the orange cases, please let it indicated.
( ) desirable ( ) indifferent ( ) no desirable
a. Do you think important to consider this scenario in an assessment procedure to be used in the presented context?
( ) yes ( ) no
b. Would you like to make any comment about the scenario (advantage, disadvantage, etc.)?
2. In your opinion, observing "Scenario-A" (Table 6), how do you consider the non-occurrence of rules obtained in RsP in RsT (cases in purple and red)? Both the cases, purple and red, represent rules obtained only in RsP, but the rules in red are extracted more than once in RsP over the groups. If needed to distinguish the purple cases of the red cases, please let it indicated.
( ) desirable ( ) indifferent ( ) no desirable
a. Do you think important to consider this scenario in an assessment procedure to be used in the presented context?
( ) yes ( ) no
b. Would you like to make any comment about the scenario (advantage, disadvantage, etc.)?

For questions "3" to "6", consider that for each rule set, RsP and RsT, it is shown only the subset related to the n most interesting rules of the domain. These subsets can be identified, for example, automatically, based on a set of objective measures – assuming that objective measures are suitable to find the most interesting knowledge of a given domain.

3. In your opinion, observing "Scenario-B" (Table 7), how do you consider the non-occurrence of some (or none) of the n most interesting rules in RsP in RsT (cases in blue)? Notice that the blue rules belong only to the RsP set.
( ) desirable ( ) indifferent ( ) no desirable

---

[7] Rule set obtained through a traditional process.
[8] Rule set obtained through a partitioned data.

Table 6. Scenario-A. This scenario was formulated based on the Sup data set described in Sect. 5. In this scenario the rules in RsP are presented in their own clusters since the aim here is to detach to the user the situations that can occur among the groups (repetitions of rules) and between RsT and RsP (occurrence/non-occurrence of rules between the sets).

RsT	RsP
CALDO_MAGGI & FERMROYAL ⇒ CREME_DE_LEITE_NESTLE	*Cluster$_n$*
PAPEL_ALUMROLITTO & GELATINA_ROYAL ⇒ FARTRIGO_RENATA	CALDO_MAGGI & FERMROYAL ⇒ CREME_DE_LEITE_NESTLE
	PAPEL_ALUMROLITTO & GELATINA_ROYAL ⇒ FARTRIGO_RENATA
	MILHO_VERDE_QUERO & FEIJAO_TORRESAN ⇒ ACHOCNESCAU
	FERMROYAL & DETERGLIMPOL ⇒ LEITE_MOCA
	Cluster$_m$
	PAPEL_ALUMROLITTO & GELATINA_ROYAL ⇒ FARTRIGO_RENATA
	FERMROYAL & DETERGLIMPOL ⇒ LEITE_MOCA

a. Do you think important to consider this scenario in an assessment procedure to be used in the presented context?

( ) yes ( ) no

b. Would you like to make any comment about the scenario (advantage, disadvantage, etc.)?

4. In your opinion, observing "Scenario-B" (Table 7), how do you consider the reverse scenario? This is, the non-occurrence of some (or none) of the n most interesting rules in RsT in RsP (cases in orange)? Notice that the orange rules belong only to the RsT set.

( ) desirable ( ) indifferent ( ) no desirable

a. Do you think important to consider this scenario in an assessment procedure to be used in the presented context?

( ) yes ( ) no

b. Would you like to make any comment about the scenario (advantage, disadvantage, etc.)?

5. In your opinion, observing "Scenario-B" (Table 7), how do you consider the existing intersection between the n most interesting rules in RsP and the n most interesting rules in RsT (cases in red)?

( ) desirable ( ) indifferent ( ) no desirable

a. Do you think important to consider this scenario in an assessment procedure to be used in the presented context?

( ) yes ( ) no

b. Would you like to make any comment about the scenario (advantage, disadvantage, etc.)?

6. In your opinion, how do you would consider the spread of the n most interesting rules in RsP in a small number of clusters?

( ) desirable ( ) indifferent ( ) no desirable

a. Do you think important to consider this scenario in an assessment procedure to be used in the presented context?

( ) yes ( ) no

b. Would you like to make any comment about the scenario (advantage, disadvantage, etc.)?

Table 7. Scenario-B. This scenario was formulated based on the Sup data set described in Sect. 5. In this scenario the rules in RsP are presented all together since, in this case, only the n most interesting rules in the set are exhibited to the user, independently of the group they were extracted – the aim here is to detach to the user the situations that can occur between the subsets containing the n most interesting rules.

RsT	RsP
n most interesting rules in RsT	*n most interesting rules in RsP*
SPO.MINERVA & BISC.NESTLE & LEITE.MOCA ⇒ ACHOCNESCAU	ACUCAR.CRISTAL.DA.BARRA & OLEO.SOJA.SOYA & SAL.CISNE ⇒ FEIJAO.TORRESAN
SHELSEVE & COCA.COLA ⇒ LEITE.MOCA	FANTA & ACUCAR.DA.BARRA & LEITE.MOCA ⇒ OLEO.SOJA.SOYA
LUSTRA.MOVPOLIFLOR & CREME.DE.LEITE.NESTLE ⇒ LEITE.MOCA	MILHO.VERDE.QUERO & FEIJAO.TORRESAN ⇒ ACHOCNESCAU
CALDO.MAGGI & FERMROYAL & LEITE.MOCA ⇒ CREME.DE.LEITE.NESTLE	SHELSEVE & COCA.COLA ⇒ LEITE.MOCA
NISSIM.LAMEM & PAPEL.HIGPERSONAL & BISC.NESTLE ⇒ ACHOCNESCAU	LUSTRA.MOVPOLIFLOR & CREME.DE.LEITE.NESTLE ⇒ LEITE.MOCA
SCLIXO.PLASLIXO & GELATINA.ROYAL ⇒ ARROZ.PRATO.FINO	CALDO.MAGGI & FERMROYAL & LEITE.MOCA ⇒ CREME.DE.LEITE.NESTLE
LUSTRA.MOVPOLIFLOR & ACUCAR.UNIAO ⇒ LEITE.MOCA	NISSIM.LAMEM & PAPEL.HIGPERSONAL & BISC.NESTLE ⇒ ACHOCNESCAU
OLEO.SOJA.SOYA & BISC.NESTLE & COCA.COLA ⇒ LEITE.MOCA	**PAPEL.ALUMROLITTO & GELATINA.ROYAL & FARTRIGO.RENATA ⇒ BISC.NESTLE**
FEIJAO.TORRESAN & PAPEL.HIGPERSONAL & BISC.NESTLE ⇒ LEITE.MOCA	SCLIXO.PLASLIXO & GELATINA.ROYAL ⇒ ARROZ.PRATO.FINO
LUSTRA.MOVPOLIFLOR & VEJA.MUSO ⇒ LEITE.MOCA	**PAPEL.ALUMROLITTO & GELATINA.ROYAL ⇒ BISC.NESTLE**
VEJA.MUSO & FERMROYAL ⇒ LEITE.MOCA	**PAPEL.ALUMROLITTO & GELATINA.ROYAL & CREME.DE.LEITE.NESTLE ⇒ FARTRIGO.RENATA**
OLEO.SOJA.SOYA & FERMROYAL & PAPEL.HIGPERSONAL ⇒ ACUCAR.DA.BARRA	OLEO.SOJA.SOYA & BISC.NESTLE & COCA.COLA ⇒ LEITE.MOCA
DESINFPINHO & CREME.DE.LEITE.NESTLE ⇒ LEITE.MOCA	SCLIXO.PLASLIXO & GELATINA.ROYAL & LEITE.MOCA ⇒ ARROZ.PRATO.FINO
PAPEL.ALUMROLITTO & GELATINA.ROYAL ⇒ FARTRIGO.RENATA	FEIJAO.TORRESAN & PAPEL.HIGPERSONAL & BISC.NESTLE ⇒ LEITE.MOCA
CALDO.KNORR & PAPEL.HIGPERSONAL ⇒ LEITE.MOCA	OLEO.SOJA.SOYA & FERMROYAL & PAPEL.HIGPERSONAL ⇒ ACUCAR.DA.BARRA
FEIJAO.TORRESAN & OLEO.SOJA.SOYA & ACHOCNESCAU ⇒ LEITE.MOCA	DESINFPINHO & CREME.DE.LEITE.NESTLE ⇒ LEITE.MOCA
CARGA.GILLETTE & BOMBRIL ⇒ LEITE.MOCA	**ESPONJA.BOMBRIL & ACHOCNESCAU & BISC.NESTLE ⇒ LEITE.MOCA**
SABPRO1EX & AGUA.SANITCANDURA ⇒ PAPEL.HIGPERSONAL	PAPEL.ALUMROLITTO & GELATINA.ROYAL ⇒ FARTRIGO.RENATA
OLEO.SOJA.SOYA & PAPEL.HIGPERSONAL & BISC.NESTLE ⇒ LEITE.MOCA	FEIJAO.TORRESAN & OLEO.SOJA.SOYA & ACHOCNESCAU ⇒ LEITE.MOCA
PAPEL.ALUMROLITTO & GELATINA.ROYAL & BISC.NESTLE ⇒ FARTRIGO.RENATA	SABPRO1EX & AGUA.SANITCANDURA ⇒ PAPEL.HIGPERSONAL
OLEO.SOJA.SOYA & FEIJAO.BROTO.LEGAL ⇒ LEITE.MOCA	OLEO.SOJA.SOYA & PAPEL.HIGPERSONAL & BISC.NESTLE ⇒ LEITE.MOCA
AGUA.SANITCANDURA & GELATINA.ROYAL ⇒ FARTRIGO.RENATA	PAPEL.ALUMROLITTO & GELATINA.ROYAL & BISC.NESTLE ⇒ FARTRIGO.RENATA
AMACCOMFORT & FARTRIGO.RENATA & CREME.DE.LEITE.NESTLE ⇒ LEITE.MOCA	OLEO.SOJA.SOYA & FEIJAO.BROTO.LEGAL ⇒ LEITE.MOCA
REQCATUPIRY & CREME.DE.LEITE.NESTLE ⇒ LEITE.MOCA	**ESPONJA.BOMBRIL & FERMROYAL & LEITE.MOCA ⇒ CREME.DE.LEITE.NESTLE**
ESPONJA.BOMBRIL & ACHOCNESCAU ⇒ LEITE.MOCA	AGUA.SANITCANDURA & GELATINA.ROYAL ⇒ FARTRIGO.RENATA
LUSTRA.MOVPOLIFLOR & BISC.NESTLE ⇒ LEITE.MOCA	AMACCOMFORT & FARTRIGO.RENATA & CREME.DE.LEITE.NESTLE ⇒ LEITE.MOCA
ESPONJA.BOMBRIL & DETERGLIMPOL & LEITE.MOCA ⇒ ACHOCNESCAU	REQCATUPIRY & CREME.DE.LEITE.NESTLE ⇒ LEITE.MOCA
CALDO.MAGGI & FERMROYAL ⇒ CREME.DE.LEITE.NESTLE	ESPONJA.BOMBRIL & ACHOCNESCAU ⇒ LEITE.MOCA
NESCAFE.TRADICAO & COCA.COLA ⇒ LEITE.MOCA	**CDCLOSEUP & ACHOCNESCAU & LEITE.MOCA ⇒ BISC.NESTLE**
	PAPEL.ALUMROLITTO & GELATINA.ROYAL & LEITE.MOCA ⇒ FARTRIGO.RENATA
	CALDO.MAGGI & FERMROYAL ⇒ CREME.DE.LEITE.NESTLE

7. In your opinion, do you consider that the amount of rules to be extracted through clustering, compared to the traditional process, should be:
( ) low ( ) average ( ) high
a. Do you think important to consider this scenario in an assessment procedure to be used in the presented context?
( ) yes ( ) no
b. Would you like to make any comment about the scenario (advantage, disadvantage, etc.)?

8. In your opinion, only in relation to RsP, do you consider that the clustering process should, as a consequence, enable each cluster to express a distinct topic of the domain?
( ) yes ( ) indifferent ( ) no
a. Do you think important to consider this scenario in an assessment procedure to be used in the presented context?
( ) yes ( ) no

b. Would you like to make any comment about the scenario (advantage, disadvantage, etc.)?

9. Can you identify other scenario(s), not previously explored, that can be relevant to the presented context? Give an example of the scenario(s) that you identified.
a. Do you think important to consider this(these) scenario(s) in an assessment procedure to be used in the presented context?
( ) yes ( ) no

10. If you want to leave any comment/observation, please do it below.

References

1. Wu, X., Kumar, V.: The Top Ten Algorithms in Data Mining. Chapman & Hall/CRC, Boca Raton (2009)
2. Dadaser-Celik, F., Celik, M., Dokuz, A.S.: Associations between stream flow and climatic variables at Kizilirmak river basin in Turkey. Glob. NEST J. **14**(3), 354–361 (2012)
3. Xiao, G.: Association rules algorithm in bank risk assessment. In: Lee, J. (ed.) Advanced Electrical and Electronics Engineering. LNEE, vol. 87, pp. 675–681. Springer, Heidelberg (2011)
4. Nuwangi, S.M., Oruthotaarachchi, C.R., Tilakaratna, J.M.P.P., Caldera, H.A.: Usage of association rules and classification techniques in knowledge extraction of diabetes. In: Proceedings of the 6th International Conference on Advanced Information Management and Service, pp. 372–377 (2010)
5. Rajasekar, U., Weng, Q.: Application of association rule mining for exploring the relationship between urban land surface temperature and biophysical/social parameters. Photogram. Eng. Remote Sens. **75**(3), 385–396 (2009)
6. Changguo, Y., Nianzhong, W., Tailei, W., Qin, Z., Xiaorong, Z.: The research on the application of association rules mining algorithm in network intrusion detection. In: Proceedings of the 1st International Workshop on Education Technology and Computer Science, vol. 2, pp. 849–852 (2009)
7. Koh, Y.S., Pears, R.: Rare association rule mining via transaction clustering. In: 7th Australasian Data Mining Conference. CRPIT, vol. 87, pp. 87–94 (2008)
8. Maquee, A., Shojaie, A.A., Mosaddar, D.: Clustering and association rules in analyzing the efficiency of maintenance system of an urban bus network. Int. J. Syst. Assur. Eng. Manage. **3**(3), 175–183 (2012)
9. Farajian, M.A., Mohammadi, S.: Mining the banking customer behavior using clustering and association rules methods. Int. J. Ind. Eng. Prod. Res. **21**(4), 239–245 (2010)
10. Fan, L.: Research on classification mining method of frequent itemset. J. Convergence Inf. Technol. **5**(8), 71–77 (2010)
11. Plasse, M., Niang, N., Saporta, G., Villeminot, A., Leblond, L.: Combined use of association rules mining and clustering methods to find relevant links between binary rare attributes in a large data set. Comput. Stat. Data Anal. **52**(1), 596–613 (2007)
12. de Carvalho, V.O., dos Santos, F.F., Rezende, S.O.: Metrics to support the evaluation of association rule clustering. In: Bellatreche, L., Mohania, M.K. (eds.) DaWaK 2013. LNCS, vol. 8057, pp. 248–259. Springer, Heidelberg (2013)
13. Aggarwal, C.C., Procopiuc, C., Yu, P.S.: Finding localized associations in market basket data. IEEE Trans. Knowl. Data Eng. **14**(1), 51–62 (2002)
14. Wang, K., Xu, C., Liu, B.: Clustering transactions using large items. In: 8th International Conference on Information and Knowledge Management, pp. 483–490 (1999)
15. Yun, C.-H., Chuang, K.-T., Chen, M.-S.: An efficient clustering algorithm for market basket data based on small large ratios. In: 25th International Computer Software and Applications Conference on Invigorating Software Development, pp. 505–510 (2001)
16. Wang, J., Karypis, G.: Summary: efficiently summarizing transactions for clustering. In: 4th IEEE International Conference on Data Mining, pp. 241–248 (2004)

17. Yang, L.: Pruning and visualizing generalized association rules in parallel coordinates. IEEE Trans. Knowl. Data Eng. **17**(1), 60–70 (2005)
18. D'Enza, A.I., Palumbo, F., Greenacre, M.: Exploratory data analysis leading towards the most interesting binary association rules. In: 11th Symposium on Applied Stochastic Models and Data Analysis, pp. 256–265 (2005)
19. Arbelaitz, O., Gurrutxaga, I., Muguerza, J., Pérez, J.M., Perona, I.: An extensive comparative study of cluster validity indices. Pattern Recogn. **46**(1), 243–256 (2013)
20. Halkidi, M., Batistakis, Y., Vazirgiannis, M.: On clustering validation techniques. J. Intell. Inf. Syst. **17**(2/3), 107–145 (2001)
21. Carvalho, V.O., Biondi, D.S., Santos, F.F., Rezende, S.O.: Labeling methods for association rule clustering. In: Proceedings of the 14th International Conference on Enterprise Information Systems, pp. 105–109 (2012)
22. Padua, R., Carvalho, V.O., Serapião, A.B.S.: Labeling association rule clustering through a genetic algorithm approach. In: Proceedings of the 17th East European Conference on Advances in Databases and Information Systems, pp. 45–52 (2013)
23. Tan, P.-N., Kumar, V., Srivastava, J.: Selecting the right objective measure for association analysis. Inf. Syst. **29**(4), 293–313 (2004)
24. Xu, R., Wunsch, D.: Clustering. Computational Intelligence. IEEE Press/Wiley, New York (2008)
25. Carvalho, V.O., Santos, F.F., Rezende, S.O., Padua, R.: PAR-COM: a new methodology for post-processing association rules. Lect. Notes Bus. Inf. Process. **102**, 66–80 (2012)
26. Carvalho, V.O., Santos, F.F., Rezende, S.O.: Post-processing association rules with clustering and objective measures. In: Proceedings of 13th International Conference on Enterprise Information Systems, vol. 1, pp. 54–63 (2011)

Author Index

Printed in the United States
By Bookmasters